The Toyota Product
Development System

The Toyota Product Development System

Integrating People, Process, and Technology

James M. Morgan
and
Jeffrey K. Liker

New York

Most Productivity Press books are available at quantity discounts when purchased in bulk. For more information, contact our Customer Service Department (888-319-5852). Address all other inquires to:

Productivity Press
444 Park Avenue South, 7th floor
New York, NY 10016
United States of America
Telephone: 212-686-5900
Fax: 212-686-5411
E-mail: info@productivitypress.com
ProductivityPress.com

Library of Congress Cataloging-in-Publication Data

Morgan, James M.
    The Toyota product development system : integrating people, process, and technology / James M. Morgan and Jeffrey K. Liker
        p. cm.
    ISBN 1-56327-282-2(alk. paper)
    1. Automobile—Design and contstruction. 2. Project management. 3. Concurrent engineering. 4. Toyota Jidosha Kogyo Kabushniki Kaisha.
I. Liker, Jeffrey K. II. Title.
    TL278.M6787 2006
    629.2'3068—dc22

                                                            2006004343

10 09 08 07 06   5 4 3 2 1

Dedication

To my wife Mary and son Greg.
Without whom this work would not be possible.

JIM MORGAN

To my wife Deb, son Jesse, and daughter Emma
You give me a loving place to come home to.

JEFF LIKER

To our colleague and friend Dr. Allen Ward
Your light still guides.

JIM MORGAN AND JEFF LIKER

CONTENTS

Contents

Contents

Contents

Contents

Contents

Contents

Contents

FIFTEEN YEARS AGO, Dan Jones, Dan Roos, and I reported in *The Machine That Changed the World* that Toyota had pioneered a new product development system. (This work was done in parallel with research by Kim Clark at the Harvard Business School and Taka Fujimoto at the University of Tokyo.) The quantitative evidence we presented was very clear: The Toyota system developed products in much less time with many fewer hours of engineering, products that cost much less to manufacture and that had many fewer defects as reported by customers. (Not surprisingly, these products also sold at considerably higher prices within a given segment of the auto market.) This product development system consistently created more value with less time and effort, the very definition of lean.

While we tried to give a broad-brush outline of how the system worked—heavy-weight program management with strong team leaders, intensive horizontal communication across departments, and simultaneous engineering—our knowledge of its details was actually very limited. After all, we were academics, not hands-on product development engineers, and our access to Toyota was limited. The best we could do was to clearly measure the difference in performance while speculating on the causes.

Surprisingly, this continued to be the situation until very recently. Everyone has understood that the Toyota system is superior, but no one has been able to describe in a comprehensive way how it actually works. And without this knowledge, efforts to copy it or even improve on it have been frustrating or impossible.

Fortunately, the volume you hold in your hand finally explains how Toyota does it and how your organization can as well. The whole gamut of Toyota methods—the chief engineer, set-based concurrent engineering (a critical concept working in parallel with simultaneous engineering), the front-loaded development process, leveled process flow, rigorous standardization of design, process, and engineering skills, etc.—is clearly explained along with the philosophy behind the use of each technique. In short, *readers working in product development organizations will no longer have any excuse for failing to copy or even to improve on the Toyota system.*

How has this breakthrough been possible? Because Jim Morgan is an actual practicing engineer—with two decades of experience in automotive product development. And he is also a scholar, having recently taken several years to do a Ph.D. at the University of Michigan. There he collaborated with Professor Jeff Liker, author of the widely praised *The Toyota Way*. Fortunately, they were able to gain extensive access to

Toyota's product development organization in the United States and Japan as part of their research.

By combining Jeff's comprehensive insights into the whole Toyota system with Jim's experience in product development, plus his fine-grained investigation of the Toyota development system, they have finally put the whole puzzle together.

All that remains is for you to study this volume carefully—and it does demand careful study because it presents a complete system integrating people, process, tools, and technology—and then to transform your own development system.

By James P. Womack

WE WOULD LIKE TO EXPRESS our deep and sincere appreciation to the many people who have contributed to this research. We are profoundly indebted to them.

We can never possibly repay the profound debt we owe to Mr. Mike Massaki, Mr. Uchi Okamoto, and Mr. Hiro Sugiura of Toyota. Massaki-san provided us with access and opportunity and Okamoto-san and Sugiura-san spent countless hours explaining the intricacies of Toyota vehicle development, while Mr. Miyadera answered many additional questions. We also gained many insights into Toyota's systems from American Toyota executives who have worked so hard to understand the true philosophy at the Toyota Technical Center, including Jim Griffith, Ed Manley, Bruce Brownlee, and David Baxter.

We also owe a considerable debt to the many others at Toyota such as Mr. Uchiyamada, Mr. M. Terasaka, Mr. S. Yamaguchi, Mr. S. Nakao, Mr. K. Miyadera Mr. T. Yamashina, Mr. E. Gay, Mr. C. Royal, Mr. T. Buffeta, Dr. C. Couch, and Mr. B. Krinock and the many others who spent their time helping us in this research.

We are of course very grateful for the work that came before this and in a sense made this book possible. The now classic work of Dr. Jim Womack and Dr. Dan Jones, as well as the value stream mapping book and class by Mr. Mike Rother and Mr. John Shook inspired and informed this application to product development. The long stream of product development research at the University of Michigan provided a firm foundation, including the excellent work by Professor Durward Sobek, Dr. Pat Hammett, Dr. John Cristiano, Dr. Jay Baron, Professor Jack Hu, and especially for the path-breaking research of Dr. Allen Ward.

We are also very grateful to those who have provided ongoing assistance and feedback in this process such as Mr. John Shook and Mr. Stephen Hung, especially with value stream mapping.

Finally, and most especially we are grateful to our families. Jim thanks his wife Mary for her editorial input, continuing moral support, and patience and son Greg for patience beyond his years, support, and for the good luck charm that he loaned. Jeff thanks his wife Deb, son Jesse, and daughter Emma who provide a loving home to return to and put up with much time away traveling to spread the word of the Toyota Way.

RESEARCH FOR THIS BOOK began in the fall of 1982 when Jeff Liker was invited to join a major comparative study of the U.S.-Japan auto industry led by David Cole and Robert Cole at the University of Michigan. The study involved most of the automakers and many suppliers in the United States and Japan and faculty across the university's departments. The study focused on the different approaches used by U.S. automakers and Japanese automakers in working with suppliers on product development. It quickly became obvious that Toyota's approach, which was distinctly different from that of U.S. automakers and only partially similar to that of Japanese automakers, was in most respects unique and exceptional, particularly with respect to its product development practices.

At that time, there was much interest in the Toyota Production System (TPS) (later referred to as "lean manufacturing") but relatively little interest in Toyota's product development system. In fact, TPS and product development had evolved quite distinctively and in separate organizational units. Most Toyota product development managers claimed they had very limited knowledge of TPS, and Toyota engineers did not see TPS as the launching point for lean processes in product development.

The research program that subsequently evolved at the University of Michigan included Al Ward, a mechanical engineering professor at the time, and a number of gifted Ph.D. students. One major theme of the study was set-based, concurrent engineering, and specifically, Al Ward's research on how Toyota engineers thought broadly about sets of solutions before zooming in on one particular solution. The study also included research conducted by Durward Sobek, whose work entailed a comparative review of Toyota's system and Chrysler's early platform team structure, detailed the chief engineer system and the mechanisms for coordination across functional specialists, and provided an applied understanding of set-based concurrent engineering. Articles based on Sobek's work were subsequently published in *Harvard Business Review* and *Sloan Management Review*.

While this research gave study participants a broad view of how Toyota engineered its vehicles and a deeper understanding of other issues, there was something missing. In some ways, we had really only skimmed the surface. What was missing was research by someone with the technical understanding to see clearly and deeply the specific and fine-grained differences between Toyota's system and that of conventional automakers and translate that understanding into actionable principles for lean implementation. The person who filled this gap was James Morgan who, over the course of two decades, had accumulated a wealth of experience in various roles in automotive product development and had served as vice president of a tier-one automotive supplier of tools, parts, and engineering services.

Over the next three years, Morgan spent many hours learning, in detail, about Toyota's body development system and about the body development system of a large North American auto company. His experience allowed him to penetrate deeply into the actual engineering processes, tools and technology employed, and people systems at Toyota and to understand the differences between Toyota and its North American foil at a detailed level. To communicate these differences Morgan utilized a sociotechnical model to compare and contrast elements of each company's people systems, processes, and technology. He also developed a value stream mapping methodology, specifically adapted to product development. This methodology later became a key tool for lean product development implementation.

In some ways, this book is the product of a dual process. To an extent, it is a product of the accumulation of over 20 years of study by the University of Michigan research group. But it is also a work heavily built on James Morgan's more recent research. In writing the book, the authors chose to present the materials that comprise the sum total of the research as a set of principles of a Lean Product Development System (LPDS). Case examples are interwoven with theory and theory is interwoven with suggestions and methodology for practical implementation of these principles. The objective is to provide companies that wish to become leaner in product development a foundation for their own LPDS.

One of the things the authors discovered while researching Toyota is how deeply Toyota's systems are rooted in the company's unique history and evolution—the Toyoda family, Japanese culture, the specific social and economic environment from which Toyota emerged and matured, and the decades of effective company-wide learning. But because every company has its own history and its own environmental circumstances, it is neither possible nor desirable for a company to adopt Toyota tools and strategies and transform itself into Toyota. It is also not possible to single out an individual tool or technique or process, change it to reflect lean principles, and expect it to operate in exactly the same way as it does elsewhere. While it is true that all companies must evolve their own systems, we hope that your journey will benefit from this research and the principles of LPDS.

James Morgan
Jeffrey Liker
Ann Arbor, Michigan

SECTION ONE

Introduction

CHAPTER 1

The New Product Development Revolution

*"There is nothing wrong with a company that
great product cannot solve."*

CARLOS GHOSN, CEO, Nissan

IN 1990, THE MACHINE THAT CHANGED THE WORLD took the automobile industry by storm, providing irrefutable proof that Japanese automakers were simply better than their European or U.S. counterparts. In fact, they were not a little better, but a lot better—two to ten times across a range of performance metrics. For many English speakers, *The Machine* was an introduction to the tremendous performance capability of the Toyota Production System. It was also an introduction to Toyota, a company destined to become an automotive juggernaut. In *The Machine,* Jim Womack, Dan Jones, and Dan Roos (1991) introduced the term *lean manufacturing*— doing more of everything with less of everything. It described a production system that was better, faster, and cheaper; required less space, less inventory, and fewer labor hours; and avoided wasteful practices. Along with subsequent works on the Toyota Production System (TPS), *The Machine* sparked a revolution in manufacturing that crossed both national and industry boundaries, spawning a multimillion dollar consulting phenomenon that has made *lean manufacturing* the most important development in manufacturing of the past two decades.

As the authors of *The Machine* are quick to point out, only one chapter of the path-breaking book focused on manufacturing. The book is really about the lean enterprise, which includes marketing, distribution, accounting, and product development. Yet most company transformation efforts have focused almost exclusively on the manufacturing shopfloor, a logical first step that more than a decade of experience implementing lean supports. But we also have learned from this experience that the shop floor is only the starting point. The transformation into a lean enterprise requires a second step: moving upstream to the development of products and processes. As many companies have discovered, there is only so much waste that you can squeeze out of production before the engineering of the products and processes becomes a critical constraint. Indeed, product and process development can have an even bigger impact on lean enterprise than lean manufacturing. Fortunately, Toyota provides as good a model for product-process development as it does for product manufacturing. Toyota's product development system, though not as well known as the Toyota Production System, is every bit as refined and powerful.

This book presents a model of a Lean Product Development System (LPDS) and is the culmination of research, experience, and insights the authors have gained over the course of many years. It is work that incorporates and integrates knowledge acquired from more than 15 years of research at the University of Michigan, more than 20 years of product development experience, privileged access to Toyota, and the patient guidance of our Toyota Sensei. It is the first research-based book to assemble together Toyota's product development practices, policies, and philosophies into one system. The research base for this book began with studies by Liker, Ward, and their students, leading to the creation of the set-based concurrent engineering model (Ward et al, 1995). Durward Sobek (1997) took this research a step forward in his dissertation through a broad comparison of Toyota's PD system to Chrysler's then emerging platform organization of product development.

Building on this research stream, Jim Morgan, while at the University of Michigan, drew on his decades of direct experience and conducted a two-and-a-half year, in-depth study of how Toyota's automotive body development compared to body development at an American "big 3 automaker." By examining one vehicle subsystem (the longest lead time and Toyotas most common type of product development), Morgan was able to penetrate deeply into Toyota's practices, which he then extrapolated into a broader model of lean product development. The scope of the study included design engineering, body engineering, manufacturing engineering, prototype development, die manufacture, body assembly development, and die and stamping approval. Data and information were gathered through interviews with Toyota and supplier representatives and during site visits. Over 1,000 hours of interviews were held with 40 people at 12 different sites in the United States and Japan. Company representatives from executive management, body engineering, manufacturing or production engineering, tool manufacture, as well as several chief engineers, participated in the interviews. Furthermore, Morgan built his study on a sociotechnical framework (people, process, technology) for analysis based on a long tradition of established research (Taylor and Felton, 1993; Nadler and Tushman, 1997).

In many ways, this book, which brings together insights from this collective research base, was inspired by a single question: What are the underlying principles of product development that have made Toyota so successful? To answer this question, the authors identified 13 principles that were subsequently grouped into three broad categories: Process, People, and Technology, and these became the framework for the Lean Product Development System (LPDS) model. The purpose of this work is to present the LPDS model in a way that demonstrates why a lean productive development system is an asset and how such a system can be created, implemented, or improved in any company. Though the discussions in this work are auto centric, the authors' experiences helping other companies implement these practices have demonstrated that the principles and processes apply to any product development system.

One of the challenges of creating the LPDS model is that the Toyota PD System is constantly evolving to meet new challenges and technologies. In fact, for the authors, the learning process has been very much the proverbial peeling away the layers of the onion, each layer revealing new and critical insights. You can describe lean manufacturing as a set of tools (e.g., *kanban, andon, poka yoke*) that eliminates waste and creates flow of materials through a transformation process. You can describe lean product development the same way. But peel away another layer, and you discover the basis of both lean product development and lean manufacturing is the *importance of appropriately integrating people, processes, tools, and technology to add value to the customer and society.*

The Next Competitive Frontier: The Product Development System

Today, lean manufacturing is no longer the exclusive competitive advantage of Toyota. Former disciples of Taiichi Ohno, the father of the Toyota Production System, have circumnavigated the globe, teaching and implementing lean manufacturing principles in many industries. In the automotive industry, lean manufacturing has become so effective that every automobile company has developed a lean manufacturing strategy, and many have been quite successful. Likewise, many companies in other industries have developed or are developing lean strategies. While most automobile companies still lack Toyota's manufacturing prowess, they have made impressive progress in closing the productivity gap; in North America, some have even surpassed Toyota in specific manufacturing categories. According to the 2005 Harbour Top Ten North American Assembly Plants hours per vehicle rating (see Figure 1-1), GM Oshawa was number one with just 15.9 hours per vehicle. In second place was Nissan's Smyrna, Tennessee plant (16.1 hours per vehicle), followed by Ford's Atlanta plant (16.6 hours); Toyota's Georgetown plant (18.4 hours); and DCX (18.7 hours). This is indeed an impressive change from the 40 hours per vehicle reported in *The Machine* for the GM plant in Framingham, Massachusetts, in the 1980s.

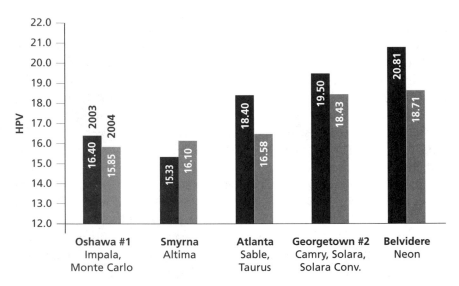

Figure 1-1. Harbour Top N.A. Assembly Plants by OEM—Hours per Vehicle

For some time, industries in the Western Hemisphere have been focusing on pushing manufacturing and routine IT activities overseas to Asia, specifically to China and India. However, the core of product-process development thinking remains the domain of the parent company, and the demands of integrating the design of complex products and processes requires even more precise coordination mechanisms when outsourcing. With the tremendous cost leverage to be gained in the design stage, it is becoming clear that the new frontier is product-process development.

In the automobile industry, the number of vehicle models available to North American consumers has increased dramatically. Conversely, the number of *unique vehicle platforms has decreased substantially.* Consequently, to be successful and remain competitive, automakers must now offer a much wider variety of vehicles, while using fewer platforms. This product-intensive environment has resulted in vehicle classes such as car/truck "crossovers," which did not exist in the late 80s but accounted for more than 16 percent of total vehicle sales in North America by 2006. Automakers are also introducing new vehicles more often. According to a Merrill Lynch analysis, new model introductions over the past five years have grown at a tremendous pace, with more than 60 new vehicles being introduced each year in the United States alone between 2003 and 2005.

In connection with this trend, many industries have been moving to platform engineering. For example, Intel recently made this a strategic priority, moving to platforms of integrated chip sets to cater to different customer segments. This strategy reflects the industry's move toward mass customization. The great success of the Centrino chip was the first model for this.

Today, customers are selecting vehicles not only on the basis of cost and quality but also on style and features. As a result, most consumer-driven companies must work to meet consumer demands by accelerating product development and bringing to market the products that customers want when they want them. Companies that miss these key market trends are being left in the dust, regardless of how efficiently they can produce yesterday's products.

Shorter technology development cycles, coupled with an explosion of new and innovative vehicle features, have put tremendous pressure on vehicle-development lead times. According to Merrill Lynch, there is a direct correlation between model age and market share. "Clearly, the older the model the lower the market share—newness wins every time."

In the late 1980s, vehicle development time from styling freeze to start of production (SOP) was typically 36 to 40 months. Today, the time required to develop a new vehicle is significantly shorter, averaging approximately 24 months. Toyota has exceeded this average by cutting development times to as low as just 15 months regularly and an incredible 10 months in a single instance.

In many companies, product-development resource growth has not kept pace with the growth of the broad range of vehicle models. In fact, the availability of an *ever-increasing* variety of vehicles has led to market microsegmentation that has critical implications for product development. Wider model variety, combined with relatively stable total sales, means that individual models are destined to have smaller total sales volumes for lower individual model sales volumes. Amortizing development costs means that development costs must be much lower than costs for traditional models destined for much higher sales volumes. For the program business equation to make sense, development costs must be much lower than in the past. Best-in-class companies have recognized and adapted themselves to this concept and it has propelled many innovations in design and tool development. In these companies, product development efficiency has already become a powerful product pricing advantage.

Because auto companies are introducing more vehicles more often, and are simultaneously facing higher quality expectations and increasing pricing pressure, they have less time to improve quality and manufacturing productivity. There is also a smaller margin of error: New vehicle introductions cannot result in a drop in vehicle quality. With shortened model life spans, companies can no longer afford a spike in defects and or a leisurely hours per vehicle pace. Instead, they must work to achieve leanproduct development and lean manufacturing that work synergistically to create flawless launches and unprecedented manufacturing quality and efficiency.

Excellence in Product Development: The Next Dominant Core Competency

Given the dramatic changes in the automotive product development environment, it is obvious that a strong product development system is a crucial core competence and fundamental to the success of any consumer-driven company. The growing complexity of the modern automobile, along with the changes discussed above, make new product development extremely challenging. In today's hyper-competitive market, excellence in product development is rapidly becoming more of a strategic differentiator than manufacturing capability. In fact, it can be argued that product development will become the dominant industry competence within the next decade.

The reason for this prediction is simple: *There is much more opportunity for competitive advantage in product development than anywhere else.* Two underlying factors support this premise. First, whereas the performance gap in manufacturing is closing, the gap between best-in-class and the rest of the automobile industry in product development is increasing. Furthermore, although most companies have made significant improvements in manufacturing since the late 1980s through the introduction of lean manufacturing methodologies, current levels of manufacturing efficiency portend that a focus on manufacturing will have diminishing returns in the future. Secondly, manufacturing's ability to impact vehicle sales performance is inherently limited. While a strong manufacturing system can affect quality and productivity, the ability to impact customer-defined value as well as vehicle investment and variable cost is clearly much greater early in the product's development process and decreases as the development program proceeds toward launch. And manufacturing can do little to reduce development costs or the timing of vehicle introduction relative to competitors, features, technology, or styling. Furthermore, manufacturing has little role in the initial selection of component suppliers. Given that most vehicles have greater than 60 percent supplier content (a common trend in other industries as well), supplier contribution to engineering and manufacturing, and, consequently, supplier selection, has a huge impact on overall vehicle cost and quality. Finally, as Toyota and others have clearly demonstrated, though manufacturing capability is paramount, it is only one functional discipline. Success requires that complementary disciplines be equally effective.

Lean Product Development System: Linking Disciplines, Departments, and Suppliers

Many progressive companies, including Toyota, are exploring opportunities to create a truly lean enterprise, not only in the manufacturing process but in design, purchasing, engineering, finance, and human resources. However, many companies continue to struggle with executing a lean enterprise strategy, partly because they have overlooked the leverage gained by linking these functions. Lean product development requires an integrated effort among sales and marketing design, purchasing, engineering, manufacturing, and suppliers. As Figure 1-2 illustrates, integrating the efforts of these disciplines in product design is fundamental to lean product development.

As previously noted, lean product development offers by far the greatest potential for a competitive advantage for any consumer-driven company and is a critical component in dealing with the many environmental challenges that all companies must now take into consideration. Many in

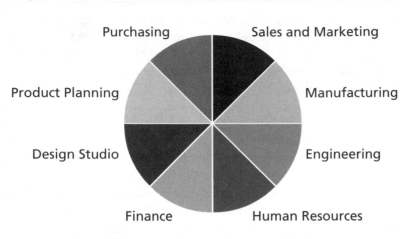

**All of these disciplines must collaborate
and execute with excellence**

Purchasing   Sales and Marketing

Product Planning   Manufacturing

Design Studio   Engineering

Finance   Human Resources

Figure 1-2. The Lean Enterprise Model

the automobile world concur with this opinion. The CEOs of GM, Ford, Nissan, and DCX have all identified product development capability as the centerpiece of their respective competitive strategies. In connection with this, these companies have taken steps to bolster product development in diverse ways, under the direction of astute individuals with the foresight and ability to implement wide-ranging changes. In the 1990s, for example, GM brought vaunted product guru Bob Lutz out of retirement and opened a giant state-of-the-art engineering campus. Ford named a new product chief and completely reorganized both its product development group and its process. DCX also named a new vice president of product development in an effort to shake up its product development system. Efforts by GM, Ford, and DCX to adopt features of Toyota's product development system have been ongoing.

Why Focus on Toyota?

It is easy to get the impression that the authors of this work and others look to Toyota for guidance in product development simply because Toyota is so good at manufacturing. But while Toyota's stunning success at consistently bringing to market excellent products, build market share globally, and make profits year in and year out is impressive, it cannot be attributed solely to lean manufacturing. The underlying philosophy of *The Toyota Way* led to parallel evolutionary paths in three core competencies—manufacturing, sales, and product development (Liker, 2004). Although not as broadly understood as Toyota's production system, Toyota's product development system is just as profound and powerful. The company consistently develops higher quality vehicles faster, for less cost, and at a greater profit than its competitors. It also launches more new vehicles annually than most of its competitors, creating a steady flow of high quality new products to meet consumer demand. This has fueled industry leading profits (reaching a Japanese record $10.9 billion by 2005 and continuing upward), a market capitalization greater than that of GM, Ford, and DCX combined, and a continuing growth in market share (on track for a world-leading 15 percent of the global market). Toyota's market value ($177 billion in 2005) exceeds the combined value of General Motors, Chrysler, and Ford and in 2005 was 13 times that of General Motors alone.

One reason for these successes is the quality of Toyota products. Objective data, showing that Toyota excels in new product quality, includes the J.D. Powers survey for initial quality from a survey of new car customers within the first 90 days of ownership. The survey is a particularly well known indicator of vehicle development quality, which Toyota has dominated during this decade with 39 first-place vehicles since 2001, including a phenomenal 10 first-place vehicles out of 16 categories rated in 2005 (see Figure 1-3).

Measure	Europe	Japan	North America	Toyota
2001 JD Powers IQS*	1	0	1	7
2002 JD Powers IQS*	0	1	2	9
2003 JD Powers IQS*	1	1	3	6
2004 JD Powers IQS*	1	0	2	7
2005 JD Powers IQS*	0	0	2	10
Freeze to SOP (mos.)	27	20	26	15
R&D $ to Sales	5.5	5.1	4.8	3.6

*number of first place vehicles.

Figure 1-3. Product Development Performance—
Latest Trends for Freeze to SOP and Research

With respect to speed to market and product freshness, Toyota can consistently bring a new body with carryover chassis and powertrain (the most common type of automotive product development) from styling freeze to start of production in just 15 months. More basic categories of vehicles, such as the *Corolla*, require only 12 months. Most of the company's competitors require 24 to 30 months to accomplish the same task. This consistent performance advantage has enabled Toyota to more than double the number of unique models the company has marketed in North America since 1990. Toyota's average vehicle age is only 1.2 years (compared to an average age of almost three years for competing automakers), and its North American product offerings have steadily increased since 1990. And this speed to market does not come at a high price. Toyota is lowest in the ratio of R&D to sales. By combining its lean production capabilities with common architecture strategies, standard processes, and component sharing, Toyota achieves an incredible overall cost advantage. It achieves speed and quality by minimizing variation and increasing the potential for predictable outcomes in an unpredictable environment.

Learning from Toyota

Our research, as well as direct experience applying the model in a number of large consumer products companies, has shown that tweaking human resource policies alone will have little sustained impact. Borrowing a tool from Toyota's lean system is equally futile as is buying the latest and greatest collaborative engineering IT system. The only way a company can make significant breakthrough improvements in product development performance is to build its own product development system with the patience and philosophical underpinnings that has led to the success of Toyota and other great companies.

One of Toyota's strengths has been the ability to learn from others, such as Ford Motor Company, quality gurus and industrial engineers from the United States, Japan, and Europe, and then carefully adapt this knowledge to its own internal systems. In doing so, Toyota has thoroughly considered the implications, piloted the new approach, studied the costs and benefits, and adapted the new approach to improve a current process. Chapter 2 of this work breaks down the result of Toyota's efforts to integrate its product development system and lays out 13 principles that make up the LPDS model.

Copious research examples have been included to assist the reader in understanding the LPDS model. In addition to examples from Toyota, we include counter-examples from one of the big three car companies, referred to as North American Car Company (NAC), which was studied intensively for comparative purposes. The comparison should help the reader understand how Toyota's product development practices differ from a more typical automotive company.

One core premise emphasized in this work is that learning by doing and institutionalizing that knowledge is the only way an organization can begin to attain the standards of excellence set by Toyota. Although each company must ultimately build a product development system that is different from Toyota's, such a system, without the underlying philosophy and management principles created and implemented by Toyota, may very well be short-lived. It is hoped that the collected wisdom and researched examples in these pages will assist you in creating and sustaining your own product development core competency, the next competitive frontier.

CHAPTER 2

The Lean Product Development System Model

*"Everything should be made as simple as possible,
but not one bit simpler."*

ALBERT EINSTEIN

IT IS INTERESTING TO SPECULATE about the secret behind Toyota's unparalleled success in the automotive industry. Did Sakichi Toyoda pilfer some ancient Samurai secret when he started the original automatic loom company? Is there a hidden army of engineers equipped with six sigma statistical packages, expert systems, and the latest cross-networked supercomputers who are really churning out Toyota's designs? The truth is actually much more straightforward. In fact, some of the American product development executives at Toyota reduce it to three words: "common sense engineering." Unfortunately, what seems like common sense to Toyota often does not seem so common outside of Toyota.

One of the reasons for the elusiveness of Toyota's secret for success is that there is no single secret pulling it all together. Toyota's success comes from hard work, excellent engineers, a culture of teamwork, an optimized process, simple but powerful tools that work, and *kaizen* that improves, improves, and improves on all of these. In short, it is a truly lean system that continually evolves.

A Sociotechnical System (STS)

Sociotechnical systems theory (STS) became popular in the 1970s and 1980s. Part of it was driven by European experiments with workplace democracy. Part was driven by American academics with engineering and social science backgrounds. When reduced to the simplest terms, STS says that to be successful an organization must find the appropriate fit between the social and technical system that fits the organizational purpose and the external environment. Broadly speaking, the technical system includes not only machines but also the policies and standard operating procedures of an organization. All of the operational policies that an industrial engineer might put in place are part of the technical system. The social system is viewed as anything having to do with the selection, development, and characteristics of an organization's people and the culture that emerges through the interaction of those people.

The term system suggests multiple interdependent parts that interact to create a complex whole. And we cannot fully understand a system simply by looking at its individual parts. Only by studying people and equipment working together can we see the way the whole functions. In addition, a system has a dynamic element—it changes over time in response to changes in the environment. The term "open-system" refers to this interaction between what is inside the organization and the outside environment.

Toyota's product development has been evolving as a living system to fit its unique environment for many years. The sociotechnical systems model used here to describe Toyota's PD System has three primary subsystems: 1) process, 2) people, and 3) tools and technology. In a lean PD system model, these three subsystems are interrelated and interdependent and affect an organization's ability to achieve its external purpose (see Figure 2-1).

Process

Mutually Supportive Aligned System Elements

Figure 2-1. Coherent Systems Approach to Product Development

STS thinking begins with three questions: 1) What is the purpose of the organization? 2) Why does it exist? and 3) What is the relevant environment of the organization? An organization can continue to exist only if it imports sufficient information and resources from the environment to sustain the system. In other words, there must be an intimate connection between the organization and its environment.

One purpose of this work is to further define the three subsystems in the STS with 13 principles that comprise the Lean Product Development System (LPDS) model. These 13 principles correspond to each of the three subsystems of the STS model as presented in Figure 2-2.

Before reviewing these 13 principles, it is important to interject a word of caution. Although it is useful for the purposes of analysis, communication, and even implementation, to decompose the Toyota Lean PD System into a LPDS model, the model does not explain the way lean product development works in reality. While a specific tool or human resource method covered in the principles may be individually valuable, what makes lean product development truly powerful is the whole system of mutually supportive tools, processes, and human systems working in harmony. Therefore, to benefit fully from this system, implementation requires a holistic systems approach that engages the entire organization. A discussion on how this works and how to maintain a systems approach to implementation is provided in the final chapters of this book. This chapter provides a brief summary of the three sociotechnical subsystems and their corresponding principles, in a structure that serves as an outline of the content of subsequent chapters.

The Process Subsystem: LPDS Principles 1 to 4

The first subsystem is *processes*, which comprises all the tasks and the sequence of tasks required to bring a product from concept to start of production. In STS terms, this is part of the technical system. In lean terms, this is what you look at when you "map the value stream" from raw material to finished goods. In an engineering process, that raw material consists of information—customer needs, past product characteristics, competitive product data, engineering principles, and other inputs that are transformed through the product development process into the complete engineering of a product that will be built by manufacturing. Virtually all companies have some type of documented process for developing products. LPDS, however, is less interested in the documented process and more interested in the actual process, the day-by-day activities by which information flows, designs evolve, tests are completed, prototypes built, and finally a finished product emerges.

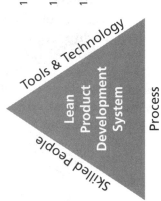

Tools & Technology

Lean Product Development System

Skilled People

Process

5. Develop a Chief Engineer System to Integrate Development from Start to Finish.

6. Organize to Balance Functional Expertise and Cross-functional Integration.

7. Develop Towering Technical Competence in all Engineers.

8. Fully Integrate Suppliers into the Product Development System.

9. Build in Learning and Continuous Improvement.

10. Build a Culture to Support Excellence and Relentless Improvement.

11. Adapt Technology to Fit your People and Process.

12. Align your Organization through Simple, Visual Communication.

13. Use Powerful Tools for Standardization and Organizational Learning.

1. Establish Customer-Defined Value to Separate Value-Added from Waste.

2. Front-Load the Product Development Process to Explore Thoroughly Alternative Solutions while there is Maximum Design Space.

3. Create a Leveled Product Development Process Flow.

4. Utilize Rigorous Standardization to Reduce Variation, and Create Flexibility and Predictable Outcomes.

Figure 2-2. Lean PPD Model and 13 Principles

Source: Reprinted with changes from Jeffrey K. Liker, *The Toyota Way* (New York: McGraw-Hill, 2004), 176, by permission of McGraw-Hill

Principle 1: Establish Customer-Defined Value to Separate Value-Added Activity from Waste

The customer is always the starting point in a lean system, so defining waste starts with defining what a customer values. Further, you must effectively communicate and operationalize customer-defined value throughout the product development organization to align all objectives, focus energy on the customer, and eliminate waste from the system. Simply put, any activity that takes time and money but does not add value from the customer's perspective is waste. In product development (PD), there are two broad categories of waste.

1. *Waste created by poor engineering that results in low levels of product or process performance.* This is the most destructive waste. The best antidote to this category of waste is a deep and concrete knowledge of how to create customer-defined value at each level of the organization, a hierarchy of value. Toyota employs tools and methods to achieve this understanding and create value and objective alignment throughout the program team.
2. *Waste in the product development process itself.* Insights from queuing theory and Product Development Value Stream Mapping (PDVSM) can help to combat these wastes.

Principle 2: Front-Load the Product Development Process While There Is Maximum Design Space to Explore Alternative Solutions Thoroughly

By far the greatest opportunity to explore alternatives is clearly early in the product development program (PD program). Toyota has developed a number of methods and techniques for effectively front-loading its product development process (PD process) with integrated cross-functional engineering resources that focus on resolving major engineering challenges while the maximum possible options are still available. Problem solving while designs are at maximum fluidity allows the company to explore potential solutions in design, engineering, and manufacturing. Furthermore, by employing this "set-based" approach (i.e., examining multiple alternatives simultaneously as opposed to single-point iteration) to engineering across functions, Toyota dramatically increases its chances of arriving at an optimal solution. This minimizes expensive engineering changes downstream. Through the discipline of *Kentou* and *Mizen Boushi* Toyota brings clarity and purpose to the front end of the PD process that drives much of the fuzziness out of the "fuzzy front end." In addition, it is able to isolate inherent PD process variation and also drives system compatibility before completing individual component designs.

Principle 3: Create a Leveled Product Development Process Flow

Once you define value and have resolved the majority of engineering and design challenges (i.e., achieved basic stability), lean product development requires a waste-free process to speed the product to market. You can manage and improve the PD process much like any other process. Although you may have many specific and unique design challenges, the tasks to be performed and their sequences are generally similar across programs. In this sense, a lean PD system is a *knowledge work job shop*, which a company can continuously improve by using adapted tools used in repetitive manufacturing processes to eliminate waste and synchronize cross-functional activities. Toyota utilizes the powerful perspective of the knowledge work job shop to level workload, create and shorten management event cadence to create a takt time, minimize queues, synchronize processes across functional departments, and reduce rework to a minimum.

Principle 4: Utilize Rigorous Standardization to Reduce Variation, and Create Flexibility and Predictable Outcomes

The challenge in product development is to reduce variation while preserving creativity. Toyota creates higher-level system flexibility by standardizing lower-level tasks. There are three broad categories of standardization at Toyota.

1. *Design standardization.* Toyota achieves this through common architecture, modularity, and reusable or shared components.
2. *Process standardization.* Toyota accomplishes this by designing products and building footprinted manufacturing facilities based on standard manufacturing processes.
3. *Engineering skill set standardization.* At Toyota, this provides flexibility in staffing and program planning.

Standardization provides the foundation for Toyota to develop effective solutions to traditionally highly cyclic resource demands inherent in most PD systems. It also allows the company to create highly stable and predictable outcomes (with both quality and timing) in an unpredictable environment.

The People Subsystem: LPDS Principles 5 to 10

The *people subsystem* covers recruiting, selecting, and training engineers, leadership style, and organizational structure and learning patterns. This subsystem and its principles cover the elusive thing called culture, which can be quite encompassing as it entails the organization's shared language, symbols, beliefs, and values. A measure of the strength of the culture, and an important tenet of lean thinking, is the degree to which an organization truly shares these things across its membership and with partners.

Principle 5: Develop a Chief Engineer System to Integrate Development from Start to Finish

In many companies, there are so many functional departments responsible for different pieces of PD that nobody is responsible. Try to identify exactly what the status of the project is or where decisions are made, and you get lost in the morass of endless departments. At Toyota, the antidote to this problem is a chief engineer who is responsible for and can tell you the exact status of any given project. The chief engineer is not just a project manager but a leader and technical systems integrator; it is to this individual that difficult decisions are brought for resolution. Although many companies have someone with the title of chief engineer or program manager, these individuals are often relegated to the role of project manager, managing people and timing but not serving as chief technical architect. The unique role of Toyota's chief engineer is to be the glue that holds the whole PD system together.

Principle 6: Organize to Balance Functional Expertise and Cross-Functional Integration

One of the more difficult tasks in developing a high-performance PD system is striking a balance between functional excellence within specific disciplines while achieving the seamless integration of those experts across departments; this synergy is required for the success of any individual program. While Toyota is fundamentally a functionally-organized company with emphasis on strong functional skills and a skill-based hierarchy, it has integrated the traditional silos through the chief engineer, module development teams, and an *obeya* ("big room") system that enhances cross-functional integration and provides a PD program focus.

Principle 7: Develop Towering Technical Competence in All Engineers

Technical excellence in engineering and design resources is fundamental to lean product development. The modern automobile is a complex system of highly technical, interdependent components that demands knowledge of computer technology, aero and fluid dynamics, mechanics, and electronics, among other specialized disciplines. In light of this, it is surprising that many automakers pay little more than lip service to developing technical superstars, preferring that their engineers broaden rather than deepen their experience and seek MBAs rather than developing their technical expertise. In fact, much of the "training" encouraged or available in many organizations is often so general as to be of questionable value at all. At Toyota, technical excellence is revered, and Toyota engineers spend a great deal of their time on core engineering. Toyota begins with a rigorous hiring process and then designs a career path that emphasizes deep technical skill acquisition within a specific discipline, focusing on mentoring of critical tactical skills that are required for engineering excellence. The principle of *genchi genbutsu* (actual part, actual place) at Toyota pushes engineers to get their hands dirty and go directly to see for themselves how the work is getting done and what the problems are.

Principle 8: Fully Integrate Suppliers into the Product Development System

Suppliers provide more than 50 percent of vehicle content for most automakers (over 75 percent in the case of Toyota) and should clearly be a fundamental part of a lean product development system. Companies should manage and nurture their suppliers in much the same way they manage and nurture internal manufacturing and engineering resources. At Toyota, suppliers are valued for their technical expertise in addition to their parts-making capability. Presourcing arrangements get them on board from the start so that they are involved from the earliest stages in concept development of a product. Using methods like having guest engineers from suppliers work full-time in Toyota's engineering offices cement the intimate relationship between Toyota and its suppliers.

Principle 9: Build in Learning and Continuous Improvement

An ability to learn and improve may well be the most sustainable competitive advantage a company has in its arsenal. At Toyota, learning and continuous improvement are a basic part of day-to-day operations.

Toyota, a leader in gathering, diffusing, and applying performance-enhancing information, recognizes the benefits of learning and maximizes its impact company-wide.

Principle 10: Build a Culture to Support Excellence and Relentless Improvement

The DNA of Toyota is a composite of very strongly held beliefs and values that are shared with successive generations of managers and working level engineers. These core beliefs compel the organization to work harmoniously toward common goals. Toyota's culture supports excellence with explicitly defined values and an unwavering adherence to core beliefs by leaders and team members alike. All of the other principles work because the culture itself makes the principles a living part of how Toyota gets things done.

The Tools and Technology Subsystem: LPDS Principles 11 to 13

The third subsystem consists of the tools and technologies employed in order to bring a vehicle into being. This subsystem not only includes CAD systems, machine technology, and digital manufacturing and testing technologies, but all the "soft" tools that support the effort of the people involved in the development project, whether it be for problem solving, learning, or standardizing best practices.

Principle 11: Adapt Technology to Fit Your People and Processes

Many companies err when they attempt to use some silver-bullet technology to achieve high levels of performance in product development, especially when they do this without considering how this technology will impact current processes or people. Adding technology to a fundamentally flawed product development system will do little to help and may even retard performance—especially for the short term. Toyota recognizes that technology in and of itself seldom represents a meaningful competitive advantage, partly because technology can be so easily replicated elsewhere. It is much more important to take the time and effort to make sure that the technology fits and enhances already optimized and disciplined processes and highly skilled and organized people. That is why Toyota works hard to adapt design software and other digital simulation tools specifically to the Toyota Way before implementing them. In an effective product development system, effective process and people subsystems come first; technological accelerators that leverage specific opportunities follow.

Principle 12: Align your Organization through Simple, Visual Communication

While culture and customer focus is the glue that holds the Toyota organization together, there are some simple tools that help align the many designers and engineers focusing on their technical specialties. One well-known Japanese management tool is *hoshin kanri*, also known as policy deployment. This is a method to break high-level corporate goals down to meaningful objectives at the working level of the organization. This method is also used by Toyota to break vehicle objectives down to specific system objectives for performance, weight, cost, safety, etc. To support this process and to solve problems that naturally occur when things do not go exactly to plan, Toyota uses very simple, visual methods for communicating information, often limiting it to one side of one sheet of paper. This A3 report (named after the A3 paper size) has four minor variations for proposals, problem solving, status up dates, and competitive analysis.

Principle 13: Use Powerful Tools for Standardization and Organizational Learning

A well-known principle of *kaizen* is that you cannot have continuous improvement without standardization. A corollary of this is that learning should extend from program to program. Toyota has created (through an evolutionary process) some powerful tools that standardize learning from program to program. This occurs at a macro-level of learning from how a design process is shared among program managers to individual lessons at the detailed technical component level captured in engineering checklists.

These then are the 13 principles that comprise our not so simple model of a Lean Product Development System. The next three sections of this book cover individual subsystems, with a chapter dedicated to each principle. At the end of the each chapter is a brief summary of LPDS basics for the principle discussed. Section V discusses how Toyota integrates these principals into a coherent system of product development and, in Chapter 17, addresses a question that transforms theory into action: How can you learn from Toyota's PD system, and develop and implement your own coherent lean product development system?

Process Subsystem

CHAPTER 3

Establish Customer-Defined Value to Separate Value-Added from Waste

"What is our business is not determined by the producer but the consumer. It is not defined by the company's name, statutes, or articles of incorporation but by the want the customer satisfies when he buys a product or service. The question can therefore only be answered by looking at the business from the outside, from the point of view of the customer."

PETER F. DRUCKER

THE PRIMARY DIRECTIVE OF ANY TRUE LEAN SYSTEM is establishing and delivering customer-defined value. However, this can be particularly difficult to do in product development. Developing the next generation of a product concept requires an accurate understanding of value from the perspective of the targeted customer. In the automobile industry, vehicle characteristics that represent value for a *RAV Four* customer and a *Lexus LS 430* sedan customer are likely to be quite different. Compounding the problem are buyer demographics (age, location, or income) and the personal preferences of individuals looking at the same type of car. Getting this right can be a very challenging task, but getting it wrong may jeopardize the success of product development efforts.

The late Al Ward believed the primary role of *effective* product development is the creation of new and profitable *value streams* for the organization. With respect to product development, this presents a twofold objective: gauging customer-defined value as accurately as possible and, based on this assessment, eliminating or reducing waste that interferes with product development that matches that value.

Waste in product development generally occurs in one of two broad areas: 1) *engineering* and 2) *product development process*. This chapter addresses the former; Chapter 5 deals with the latter. In each case, the underlying premise is that no effort or resources should be expended without a deep understanding of customer-defined value.

Understanding customer preference is a basic part of any product development system. Traditional product development uses many tools to achieve this, such as market data, focus groups, and surveys. Each of these traditional tools can gather important information about market trends and customer sentiment, but they do not provide the deep knowledge that establishes a concrete understanding of customer-defined value. Because a lean PD system relies on this deep knowledge, these traditional tools fall short of the intended mark because they fail to reflect data that allows you to differentiate between value-added and non-value-added activities. Without this knowledge, it is not possible to define a customer's *value characteristics* accurately. This will adversely affect effective allocation and management of product development resources. The North American Car Company (NAC) case study below should provide a better understanding of the relationship between a traditional approach to product development and detrimental engineering waste.

Customer-Defined Value Process at North American Car Company

Every company spends a great deal of time trying to identify product attributes that will be important to customers. The North American Car Company (NAC) spends significant resources gathering and studying demographic data, reviewing the results of focus groups, benchmarking competitors, and reviewing field quality data on the current model vehicle. Once compiled, this data produce a long and detailed document describing the target customer and feasible cost model, as well as outlining *vehicle-level performance objectives* for new models. NAC utilizes sophisticated analytical tools and scores of business case reviews to evaluate objective data that determines program direction and feasibility. But something is elementally lacking. First of all, at least part of the problem is that this analytic approach may be too objective. The primary focus of the product development team is on delivering the numbers—especially the financial requirements. In fact, the teams with which we met obsessed over little else as they prepared to navigate their way through senior management reviews. Obviously this does little to establish an emotional customer connection or a shared sense of excitement about the vehicle program. The targeted customer is not central to the process—the numbers are. In fact, beyond the conceptual stage the customer is rarely even mentioned. So although the key program leadership at NAC is clearly mindful of Return on Investments (ROI), program leaders seem to be out of touch with the customer, do not focus on the customer's true value characteristics, and are unaware of the engineering waste created as a result of this.

Like other traditional companies, NAC also struggles to communicate what it knows about the program in meaningful or measurable ways to the program team, causing the team to be uncertain about the program objectives, functional objectives, and team goals. In fact, many key program participants have only a vague knowledge of their own function-specific objectives and an even vaguer understanding of general program objectives.

This lack of understanding is even more evident downstream, where groups are likely to have their own set of objectives for the vehicle program based on issues that relate to day-to-day manufacturing operations. The reason for this is simple. NAC does not involve functional groups, such as manufacturing, in the value-definition process. As a result, the NAC program objectives do not gain the full enrollment or ownership by functional groups downstream, so these groups do not have an opportunity to put into operation vehicle-level objectives or deliver on specific goals in a way that is meaningful to them. The result of failing to involve and align all program participants means each functional group develops its own goals, causing confusion or conflict across development teams. This not only inhibits NAC's ability to deliver customer-defined value but also creates program delays, significant cost penalties, and a generally inferior product.

Customer-Defined Value Process at Toyota

Like NAC, Toyota evaluates field quality data, market research, and competitor's products to understand its customer. But the similarities end here. A critical first step in Toyota's value discovery process is selecting key program leadership. Toyota selects program leaders with the background and experience to establish an emotional connection with the target customer. As Kousuke Shiramizu, *Lexus* quality guru and executive vice president explains, "Engineers who have never set foot in Beverly Hills have no business designing a *Lexus*. Nor has anybody who has never experienced driving on the Autobahn firsthand."

Program Leadership: The Chief Engineer Role

The top program leadership at Toyota is the chief engineer (CE). In addition to being a super engineer, he or she must understand what customers value and how these value characteristics mesh with the program's vehicle performance characteristics. Toyota's chief engineers and their staffs (CE team) go to great lengths to achieve this understanding. One anecdotal example of how far a Toyota CE will go to make this happen illustrates this diligence. The story concerns a chief engineer who moved in with a young target family in southern California to enhance his understanding of the generation X lifestyle associated with *RAV Four* customers. While developing Toyota's successful 2003 *Sienna*, the Sienna CE drove his team in Toyota's previous minivan model more than 50,000 miles across North America through every part of Canada, the United States, and Mexico. The CE experienced a visceral lesson in what is important to the North American minivan driver and discovered in every locale new opportunities for improving the current product. As a result, the *Sienna* was made big enough to hold full sheets of plywood while the turning radius was tightened, more cupholders were added, and cross-wind stability was enhanced, among many other improvements that resulted from this experience.

The CE's actions confirm that Toyota's *value targeting process* and *vehicle drive analysis* are important tools in the company's search for customer value. To assure that the driving experience achieves maximum benefit, CE team members receive advanced driving training as well as vehicle evaluation-skill training to identify problems and recognize improvement opportunities.

Once defined, the value characteristics must be: 1) communicated across the program to all product teams and 2) aligned and put into operation with meaningful, measurable objectives that entail specific tasks that each person on the product team can execute.

Steps for Delivering Value to the Customer

At Toyota, the chief engineer's ultimate responsibility is delivering value to the customer, though the process entails many steps and people. First, the CE communicates customer-defined value, vehicle-level performance objectives, and aligns the vehicle-level performance goals of the entire program team. This step begins with the *Chief Engineer's Concept Paper*, which outlines the CE's vision for the new vehicle. The concept paper, a document that rarely exceeds 25 pages, usually takes several months to complete. It includes both quantitative and qualitative objectives for vehicle characteristics, performance, cost, and quality. Many people provide input for the concept paper but it is written and issued by the CE and finally presented in a large auditorium as the marching orders for all participants.

The Japanese term for the concept paper (and other documents issued by the CE) is *shijisho*, literally translated as "direct order document."[1] The CE's concept paper can thus be compared to a direct order in the military. The concept is the result of many months of discussion, information gathering, and consensus building, and has been approved by the managing directors of the company, but, when it is issued, it is the law of the program.

Once the concept is approved, the next step in the customer-defined value process is to develop specific objectives that support the chief engineer's vision for all *functional program teams*. The *vehicle-level performance* goals set by the CE must be translated into specific, measurable objectives for the stylists, packaging engineers, body engineers, stamping engineers, etc. that make up the program team. Putting into operation customer-defined value at a vehicle level that supports aligned goals that cascade throughout the program teams creates a *value hierarchy*. As the CE team moves down this value hierarchy, it decomposes the high-level vehicle-level performance targets and aligns them at each level into a set of specific actions. This process gives Toyota an internal customer perspective for each functional team. (A simple example of the *value decomposition process*, which is a semistructured, reductionist approach to achieving value alignment in the product development process is provided later in the chapter.)

Next, *module development teams* (MDT), responsible for each vehicle subsystem, meet to develop specific, measurable goals for each subsystem and communicate it to the CE team. Using a customer-first attitude and the CE as the primary voice of the customer, the various MDTs go through fairly intense negotiations and ultimately commit to specific objectives designed to support the vehicle-level performance characteristics. This process is very similar to the *catch-ball* process utilized in *Hoshin Kanri*, which is a process designed to achieve enrollment and alignment among all participants and to encourage a candid exchange of views that set achievable stretch goals. This process drives everyone to focus all efforts and energy toward delivering value to the customer and guards against the

---

1. There are two types of *shijisho*. A-Types are broad and set general goals and objectives, identify trends, and include targets for weight, performance, and cost like the concept paper. B-Types are at the smaller component level, e.g., deciding the number of prototypes or, for engines, deciding on a certain type of mount.

very real potential for team members to champion their respective individual systems at the expense of optimizing the whole system. The final version of these objectives is posted and tracked throughout the program. Team members' performance is judged, in part, by their ability to hit these targets. This results in each member of the program contributing directly to delivering customer-defined value.

The next step in the process requires intensive cross-functional participation among the MDT to develop specific strategies and *value targets* to deliver the value-driven commitments each team made. Equipped with predetermined value targets, the various MDTs work together by studying field quality data, tearing down competitor products, and visiting dealerships to document direct customer feedback. They also visit their own and competitor's manufacturing plants to study production as well as talk with operators about manufacturing quality and efficiency. This is an example of an important concept described in *The Toyota Way*, *Genchi Genbutsu* (go to the source), which is a critical underpinning of the lean product development system.

It is important to understand that as the cross-functional MDTs go to the source, they are going with a common set of objectives and goals based on vehicle-level performance objectives set by the chief engineer. Because the MDTs begin their quest for delivering value early in the process, while the vehicle concept is most fluid, they are able to communicate and integrate their value-driven commitments with the design, engineering, processing, and manufacturing departments, which presents many opportunities to discover potential improvements to their development ideas. (This topic is revisited in the next chapter.)

Case Example: *Lexus* Body Team Reduces the Margin for Error in Half

A look at the early days of *Lexus* provides a simple illustration of value alignment and the value decomposition process. The driving principle behind the *Lexus* brand was *the relentless pursuit of perfection*. At the time, however, Toyota's customer feedback clearly indicated customer preference for the precision and craftsmanship offered by German engineered BMWs and Mercedes, two of the primary luxury car competitor targets established by the *Lexus* team. To achieve breakthrough quality to fit the level of quality customers would expect of a *Lexus*, the team established the high-level objective of reducing its margin of error by half. One specific action resulting from this objective was that the body team reduced the margins or small spaces between body panels from the current standard gap and also reduced the amount of acceptable variation in those gaps to half of the then accepted level, consequently achieving a level of precision not only never before seen in a vehicle body, but one considered by many at that time to be unobtainable.

The *Lexus* team realized that reducing these gaps would contribute to vehicle aerodynamics and a significant reduction in wind noise. In addition, the modifications would dramatically improve vehicle appearance and craftsmanship, something that customers had identified as being critical to selecting a luxury brand. The module development team that was responsible for vehicle closure subassemblies (e.g., hood, doors and deck lid) had a particularly difficult challenge. In addition to the already challenging goal of reducing gaps and variation tolerance, they had to consider the swing gap required for closing and opening these subassemblies. The cross-functional team began by executing a detailed teardown and analysis of two best-in-class competitors' doors. They identified very specific areas of concern, plotted on the radar graph in Figure 3-1.

The areas identified on the radar graph correspond to the specific areas on the doors shown in Figure 3-2. Since these areas showed the greatest difference between the products, the team began its investigation by analyzing these specific areas.

As Figure 3-2 illustrates, the top and bottom sections of the doors have comparable variation in gap. The greatest differences between the two doors are in area A, at the front edge of the door, and area E, at the rear edge of the door. During the manufacturing process, the door edge is created by hemming or folding the edge flange of the outer door over the periphery trim flange of the inner door as shown in Figure 3-3 on page 35.

Although the general appearance or design effects of the doors' front edges appear similar, closer examination by the team identified several important differences. The sharp break created by a drastic change of direction at point A of the *Lexus* model may create material compression and distortion in this area of the hem and subsequent variation in the panel gap. The competitor's door edge gently sweeps upward, creating a smooth gradual effect that minimizes material distortion. In the same area, they also saw that the character or style line that crosses the door interrupts the edge, creating potential for even more distortion. The style line in the competitor's door blends upward and washes out in the mirror attach area, never interrupting the hem edge.

Front Door Margin Comparison

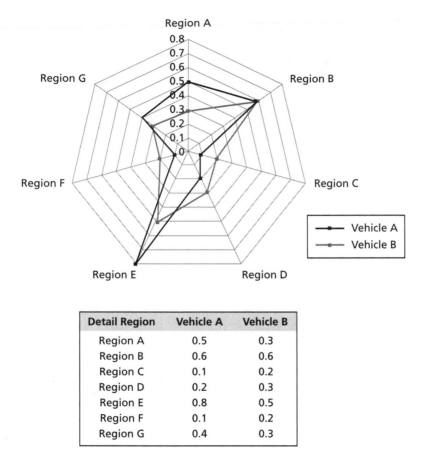

Detail Region	Vehicle A	Vehicle B
Region A	0.5	0.3
Region B	0.6	0.6
Region C	0.1	0.2
Region D	0.2	0.3
Region E	0.8	0.5
Region F	0.1	0.2
Region G	0.4	0.3

Figure 3-1. Radar Graph Comparison for Front Doors

On the rear edge of the door, the MDT discovered similar conditions: two style lines that interrupt the hem edges and another sharp break at the top of the door. With the information gleaned from this analysis, the team was able to design changes to help achieve their panel gap objectives. The team also noticed a 1mm offset in the edge of the inner door flange that follows the entire periphery (area E). Team members concluded that this offset would provide more stability than a flat edge, contributing to both gap and flushness consistency and a generally more precise panel fit

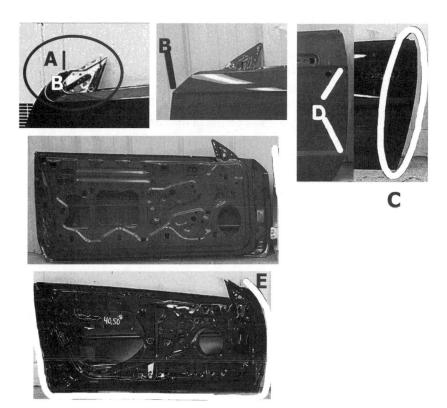

A. Front edge of door blended into mirror detail
B. Styling line integrated into mirror detail to avoid violating hem line
C. Cleaner periphery line/arc on rear edge of door
D. Sharp break-lines through hem on Vehicle A
E. Product details on inner door closely follows periphery of hem flange on Vehicle B. This constant offset provides superior dimensional stability for a better fit.

Figure 3-2. Comparison Based on Radar Graph Results

Figure 3-3. Hemming Operation

condition. This design change was also easily incorporated at this point in the program.

The MDT next visited the manufacturing facility responsible for hemming current door subassemblies. Firsthand observation and operator interviews identified another potential improvement opportunity. By changing the preclinch condition and shortening the flange lengths on the outer panels, they could achieve much greater control in the hemming process, resulting in less variation and the ability to make the gaps between panels (margin for error) smaller. Subsequent trials, in the manufacturing environment gave the team the opportunity to identify the optimum flange length and preclinch condition for both maximum clinching power (holding the assemblies together) and minimum, most constant gap.

Hem variation and panel gap and flushness were monitored throughout the program and checked closely at all design reviews; results were tracked. The team also developed specific metrics for both the end condition (4mm gap and variation) and the means (design changes, flange length, and preclinch requirements) and tracked both the means and the results throughout the program, posting simple graphs in the program team room. By tracking both the results and the means, the team was able to identify quickly whether the methods/means were resulting in suboptimal results or whether the methods/means were not being followed. By working collaboratively on the details of delivering customer value, the team was able to deliver a luxury product that exceeded standards of the time. Many readers may recall the resulting *Lexus* commercial where a ball bearing is rolled down one of the panel gaps to demonstrate a new level of vehicle precision achieved by the *Lexus* team.

Excellent product development requires that the program leadership has a process for clearly communicating specific, detailed goals that are aligned throughout the program and that leadership engages all functional groups (development teams) to participate in delivering customer-defined value. Committing to and delivering on goals early will ensure meaningful participation. It will also eliminate engineering waste, enabling manufacturers to deliver value consistently.

Why This Is the First Principle

This LPDS principle emphasizes how essential it is to a lean PD system to establish a deep understanding of customer-defined value, diffuse it throughout the program team, and decompose it into meaningful objectives throughout the program. *Establishing and decomposing customer-defined value is your critical first step in creating a lean PD process.* Furthermore, to bring this high-level knowledge about and have customer-defined value bear directly on the product development system, you need a chief engineer, or equivalent, to define and align program objectives and goals among cross-functional teams early in the program. For Toyota, giving the module development team time to analyze, track, and consider options is crucial for delivering on the performance objectives and performance goals set by the CE team.

Furthermore, when you work to establish customer-defined value, you are also separating your value-added activities from wasteful activities. It is here, at the front end of the PD process, that companies have the greatest opportunities to have a positive impact on end product. Like lean manufacturing, lean product process needs to drive the principle of eliminating nonvalue-added activities, in this case, engineering waste.

Chapter 5, which addresses LPDS Principle 3, includes a discussion of how Toyota minimizes the second PD waste, product development process waste, and creates flow by maximizing the potential of the early stages of product development. Toyota accomplishes this by front-loading the lean PD process with experienced, cross-functional engineering resources focused on resolving key engineering challenges and problem solving, the subject of the second LPDS principle.

LPDS Basics for Principle One

Establish customer-defined value to separate value-added activity from waste in product development. A lean product development system starts with the customer. There must be a process for identifying product-specific, customer-defined value, effectively communicating that value, and developing and executing specific, aligned objectives throughout the organization from the start of the program. Toyota starts with direct, visceral product and customer experiences for program leadership and then adds rigorous internal and competitor data analysis and broad technology reviews. The CE then communicates his vision for the vehicle through the CE paper and aligned, executable objectives are developed by each of the subsystem teams and confirmed by the CE. These objectives are tracked throughout the program.

CHAPTER 4

Front-Load the PD Process to Explore Alternatives Thoroughly

"The manager's job is to prevent decisions from being made too quickly ... but once a decision is made, we change it only if absolutely necessary."

TOYOTA GENERAL MANAGER OF BODY ENGINEERING

THE ABILITY TO INFLUENCE THE SUCCESS of a PD program is never greater than at the start of a project. The further into the process, the greater the constraints on decision making. As the program progresses, the design space fills, investments are made, and changing course becomes increasingly more expensive, time consuming, and detrimental to product integrity.

Empirical evidence shows that poor decisions early in the process have a negative impact on cost and timing, which increases exponentially as time passes and the project matures. Although this is generally recognized, very few companies understand how to take advantage of this golden front-end opportunity by making wise front-end investments. Toyota is one of the few.

The LPDS model's second principle, *Front-Loading the Product Development Process to Explore Alternatives Thoroughly,* is the foundation for flawless execution throughout the program and is analogous to the shopfloor maxim of *measure twice—cut once.* It emphasizes the value of good preparation and concentrates the PD team's effort on the beginning of the program. Moreover, although initial plans are often subject to change, *the process of detailed, rigorous planning is absolutely crucial to the success of the lean PD program.* It is working though the planning process; bringing together your brightest, most experienced engineers from all functional disciplines to work collaboratively, thoroughly thinking though all of the critical project details, anticipating problems, applying lessons learned, creating precise plans, and designing countermeasures from a total systems perspective that is crucial to the success of the program and the evolution of the lean product development system.

Toyota's maxim for product development planning is "plan carefully and execute exactly," and it is through rigorous planning that Toyota brings a unique precision to the product development process. It is at that stage of the PD program that a company must use its best resources for a disciplined front-loading strategy.

Because front-loading solves problems at a root cause level early in the process, it nearly eliminates the traditional PD problem of late design changes, which are expensive, suboptimal and always degrade both product and process performance. Late engineering changes are "quick fixes," or patches, not continuous improvement; in fact, they are waste of the very worst kind. In lean thinking, effective continuous improvement activity begins at the beginning of a program. Front-loading also provides a way to control much of the variation inherent in the product development process. Since early variation has the greatest impact on queues and other systems delays during the execution phase of product development, Toyota tries to isolate and minimize variation early in the process in two ways:

1. By standardizing architecture, processes, very specific activities and by setting very specific performance goals.
2. By creating an early phase of the PD process (*kentou*) to solve problems, resolve conflict, address the roots of variation, and segregate it from the rest of the PD process. This allows participants to focus on the execution of their specific tasks.

The ensuing discussion of front-loading has been subdivided into two broad categories, the first of which addresses cross-program front-loading. This section explains how Toyota "front-loads" the multiprogram "design factory" and creates an operating environment in which each individual program has the greatest potential for success. This section also discusses how Toyota leverages platforms and shared architectures, re-uses components, allocates shared resources, and assesses new technologies to minimize variation and uncertainty in product development. The second section illustrates how front-loading affects individual programs and examines what Toyota calls the *kentou* or study phase of a program as well as critical concepts like *Mizen Boushi* or designed-in quality. This chapter also covers set-based concurrent engineering, which considers sets of design and manufacturing solutions concurrently and then gradually narrows the sets, helping to ensure that designs are compatible with their environment and feasible. Set-based concurrent engineering dramatically reduces the need for engineering changes. This set-based philosophy also helps to identify and resolve problems as early as possible and ensures that product attributes, including crucial trade-offs, which are a fundamental part of product development, are clearly understood. The chapter concludes with a discussion of the principle of right person, right work, and right time as an antidote to the tendency of some companies to do too much too soon thereby creating tremendous waste. This is the waste of overproduction and creates massive rework downstream in the process because of work performed too early in a process.

Front-Loading for the Design Factory: Creating the Context for Individual Program Development by Managing Product Platforms

Any individual PD project represents only a fraction of a company's product portfolio and a single element in its total product development strategy. To be successful, an enterprise must effectively master what Cusumano and Nobeoka (1998) refer to as *multiproject management*, a strategy that optimizes the sharing of resources across multiple, concurrent projects. It is an effective way to manage technological complexities when developing diverse and sophisticated products.

A product development project can range from programs related to breakthrough inventions the world has never before seen to routine upgrades of existing products. There are four broad categories of new product development:

1. *Revolutionary new products that represent radically different products or technology.* This is a complete, white sheet product, and is the least common type of product development project in the automotive industry or any other mature industry.
2. *Product platform-development projects that require fundamentally new systems and components.* In the automotive industry, this would include such things as a new engine, transmission, chassis, underbody, HVAC, and electrical systems, resulting in a completely new vehicle that utilizes improved versions of existing technology, possibly combined in innovative ways. This is also fairly rare, particularly at Toyota, which has worked hard to maintain common "platforms" for vehicle derivatives.
3. *Derivative products built on existing product platforms.* In the auto industry, derivative vehicles can require entirely new exterior body shells, interiors, secondary technology synthesis, and trim and

ornamentation. This type of product, increasingly more common to the industry, is the primary focus of this book.

4. *Incremental product improvements.* In the auto industry, this would include such things as new trim, cladding, replacing selected outer and/or interior panels, or updating selected technologies. This "facelift"-type product development was once adequate for between-model updates, but increased competition, faster technological development cycles, and better-informed consumers are making it a less viable option, both in the automobile world and in many platform-based industries.

This section and the chapters to follow will focus on the third type of product development and Toyota's adherence to the *front-load the PD process and thoroughly explore alternatives* principle.

Derivative Vehicles Built on Existing Product Platforms

In this category of product development, standard manufacturing processes, robust common platforms, and shared vehicle architecture are crucial PD system enablers for individual programs. They contribute directly to development speed, lower development costs, and significantly higher vehicle quality. Although definitions vary, it is generally accepted that a vehicle platform contains at least the following components: power pack (engine and transmission); front, center pan, and rear end structures, front and rear axles and suspensions; frames and subframes; brake and electrical systems; bumper beams; and fuel tank. These components make up the mechanical system or "guts" of the vehicle. Much of the expensive high technology, as well as the basic functionality of the car, truck, or SUV is driven by these components. To a great extent, these components also determine vehicle driving performance characteristics and basic vehicle reliability. Common front-end structures also contribute to similar crash paths and minimize iterative testing requirements. In terms of lean enterprise, then, it makes sense to engineer standard vehicle platforms that have a great impact on vehicle quality and reliability and can accommodate a wide variety of body shells and interiors. While such platforms and the philosophy that supports them are invisible to consumers, they are the primary source of vehicle quality and reliability and an important source of differentiation that appeals to customers, a connection that Toyota has fully understood and used to refine its PD.

One example of this is the way Toyota addresses the issue of noise and vibration, a critical point among "customer value" elements. The way the engine is mounted to the steel sheet metal cradle affects the amount of noise and vibration the customer experiences, so this in particular defines a platform from Toyota's perspective. Once Toyota tests and proves a platform, the platform can be stretched and widened with a great degree of flexibility to accommodate model derivatives. For example, the Toyota *Camry, Sienna Minivan,* and *Avalon,* even though they are different sizes and appear to be very different vehicles, are all assembled on the same platform. On this platform, however, everything from the underbody and power pack up to sheet metal and interior is customized for each vehicle. In fact, they do not share a single common piece of sheet metal.

One reason Toyota stands out in the industry in terms of the number of vehicles derived from a single platform is that it develops vehicle platforms that are meant to be reused for up to 15 years. On average, Toyota produces seven different vehicles on each platform with a high degree of mechanical reliability across vehicle type. To enable "off the shelf" platform selection by individual programs, Toyota focuses on engineering reliable, robust vehicle platforms with maximum flexibility, *in advance of specific vehicle programs.* This front-loading practice is a key to Toyota's reputation for vehicle safety and reliability. It also dramatically reduces product development time and cost, in some cases even eliminating the need for prototype tools.

In addition to sharing mechanical platforms, Toyota focuses on commonizing certain critical aspects of vehicle geometry. More specifically, certain shapes, forms, and holes in exterior-stamped sheet metal and subassemblies necessary for efficient manufacturing (or successful vehicle crash performance) are identified and standardized across certain specific vehicle models or generations of the same vehicle. Examples of this include crash-critical geometry in inner hoods and doors, specific flats and holes for locating detail parts at assembly for flexible body shops, maximum hits per part, and depth-to-width ratios for parts.

Furthermore, Toyota defines standards that minimize impact on creativity by focusing on unseen part characteristics and utilizing ratios and *trade-off curves* wherever possible, allowing its designers to maintain maximum flexibility in decision making. These standards are the subject of numerous documents that will be discussed in some detail in Chapters 6 and 15. Here, it is important to emphasize that the information in these documents is available to all program participants and is an essential tool for sharing the experience and knowledge gained from other programs, thus drastically shortening technical learning curves for individual programs. It also frees up time and resources, allowing PD teams to perform other tasks and thoroughly explore alternatives during front-loading.

Advanced Technology Planning

While platforms and shared architecture provide an important foundation, the essence of new product development is delivering fresh, innovative products that excite and motivate customers to spend their hard-earned cash. As previously noted, too much commonality can lead to poor product differentiation and an uninspired portfolio of products that age on showroom floors. A lean product development system must balance quality, speed, and cost advantages of commonality with attractiveness and excitement. By combining and balancing these elements, a lean enterprise keeps a steady stream of innovative products flowing to a steady stream of customers. This requires advanced technological planning, one of the key factors that make innovation feasible.

Toyota draws from many sources for new technology and innovation: internal research and development, suppliers, Business Revolution Teams, and even competitors. All vehicle lines review R&D proposals on a regular basis, and specific R&D projects are generated from specific vehicle line input based on customer feedback and environmental changes. All primary suppliers participate in routine technology reviews; indeed, participation is understood as a prerequisite for any supplier that wants to remain a primary supplier. Business revolution teams are small, fast-moving units, which focus on technological innovation to address specific existing or anticipated challenges and feedback. One example of the efficacy of such units is the Toyota *Prius*, which started as a business revolution team project tasked with coming up with a new vehicle concept and new product development process for the twenty-first century (see Chapter 7). Finally, Toyota regularly analyzes competitor innovations for potential fit with Toyota vehicle line strategies. Teams review and subject the ideas from all these sources to a rigorous funneling process to evaluate and select concepts based on fit with strategic objectives and vehicle line requirements (i.e., ability to share across individual vehicle models).

Toyota has developed a time-based process for evaluating input, selecting ideas, developing concepts, and launching advanced projects on an annual cadence, thus assuring a steady and focused supply of innovation to each of its vehicle lines. During this process, which supplements the actual development process for specific vehicles, Toyota evaluates ideas and technologies for implementation readiness and fit with corporate requirements, including design, manufacturing, marketing, and product planning. Each new technology must pass rigorous tests before it is deemed suitable for inclusion in a specific vehicle program. Toyota is somewhat technically conservative and insists that all technologies and concepts be fully vetted before they are adopted. The company does not normally develop new technology on individual development program critical paths and is extremely demanding of suppliers who promote new technologies, requiring that all such promotions be backed by data from the start. In this way, Toyota creates a set of proven technologies that are put "on the shelf" until they are needed for specific vehicle programs. The Chief Engineers decide when and whether to pull these technologies off the shelf because it is the chief engineers who thoroughly understand what customers want and how the total vehicle fits together.

The platform planning approach, standardization, and the speed of Toyota's process enable regular insertion of technology into vehicles. If a particular technology is not thoroughly tested and reviewed by the time a chief engineer develops the concept for a vehicle, it may be incorporated into the next program or the one following it. The short cycles between vehicles ensures that the next available program will not be far behind.

Toyota's chief engineers have also developed an intuitive "feel" for how much change in a particular vehicle is just enough. The primary intent is to carry over most of the parts of the vehicle and consider the best utilization of existing tooling; only then will a chief engineer consider where and how to introduce new technologies. This is a stark contrast to the "clean sheet" approach historically employed by NAC and other companies.

It is also important to comment briefly on Toyota from an organizational perspective beyond the role of the chief engineers who decide when and for what purpose new technology is pulled from the shelf. Toyota also has a contingent of Advanced Engineers, a group that functions as an autonomous entity but is tightly linked with specific Vehicle Centers. Advanced Engineers are not pulled in to day-to-day activities. Instead, they are given the independence to create the brand identification that is so important to and necessary for a full understanding of technological fit.

One may get the impression that the high hurdles for new technology make Toyota inflexible and behind the times. Quite the contrary, technological innovation is strategically focused, often in response to a request from a chief engineer. Also, certain car models have been selected as the leaders in new technology, in particular the *Crown* model in Japan and *Lexus*. These product lines must be at the forefront of new technology in at least several areas and there is a close interaction between the research group and the CE. These technologies will, over time, migrate to less expensive models. This interaction between those developing the technologies and those leading the programs and the rigorous evaluation process before putting the technology on the shelf ensures that there is little delay between when a technology is ready and when a chief engineer determines that a program wants the technology.

Front-Loading Within an Individual Program: Styling and Engineering Feasibility

Vehicle styling, the activity that defines the exterior appearance of a vehicle, is arguably the most creative or artistic set of activities in the PD process. It is crucial to customer appeal and vehicle sales. Vehicle style also has an important impact on downstream engineering and manufacturing activities and must strive to balance its own activities with these. Styling, in any PD system, represents the joining of art and science.

At Toyota, styling studios in Europe, Asia, and North America compete with each other to capture the essence of a new car's style while incorporating sound engineering principles and guidelines set by the Toyota Production System. The styling process begins with a series of sketches, and progresses to multiple scaled-down clay models. The goal of each of these activities is to capture and refine the chief engineer's vision for the new vehicle. Styling for an existing vehicle line (by far the most common PD projects in the automotive industry), usually requires some combination of creating fresh styling queues and maintaining a given vehicle's brand DNA—retaining those styling features and vehicle characteristics that clearly identify that specific vehicle over time. Eventually the PD team creates between two to four full-scale clay models of the vehicle. During this process, the CE Team continues the practice of *genchi gembutsu*, frequently visiting the studios and sharing relevant data and analysis as it becomes available. At the same time, the CE is soliciting and receiving broad input, mostly from internal employees to guard confidentiality. The input requested and received is related to the strengths and weaknesses of alternative models.

Set-Based Concurrent Engineering

Examining multiple alternatives during the styling activity is an example of *set-based concurrent engineering* (Ward, Liker, Sobek, Cristiano, 1995). The term was coined by the authors cited here (an academic team from the University of Michigan) who studied Toyota's process and noted the unusual degree to which Toyota considers a broad range of alternatives and systematically narrows the sets to a final, often superior, choice. Morgan (2002) subsequently confirmed the set-based approach to body engineering was employed by Toyota during the *Kentou* period when hundreds of *kentouzu* (study drawings) are generated and broadly considered and matured until converging on a single design solution.

A simple graphic can help illustrate set-based concurrent engineering (SBCE) (see Figure 4-1). While conducting the early studies, the team from the University of Michigan interviewed many engineers working for U.S. and Japanese automakers, including engineers from Toyota. During these interviews, a common theme among these engineers (with the exception of Toyota engineers) was some version of "iterative point-based design." As a rule, this meant that styling would initially consider many alternatives. After these alternatives were narrowed down, executives would select a clay model and then authorize engineering to turn it into

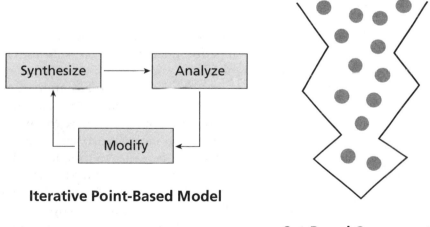

Iterative Point-Based Model

Set-Based Concurrent Engineering

Figure 4-1. Two Models of Design

digital data and a tooled factory. Of course, engineering would find many problems with the style from the perspective of functionality (e.g., aerodynamics) and request changes—iteration. This iteration would go on until time in the PD process ran out. Sometimes engineering would simply modify some aspect of the styling just to "make it work." Then, process engineering would start the work of tooling and discover further problems with the design, leading to more iterative discussion that sometimes resulted in heated conflict. It is easy to see how this iterative cycling consumes a company's time and resources and leads to a suboptimal design.

From Toyota engineers and subsequent studies of Toyota's PD systems, the team ascertained that the company was not constrained by the "iterative point-based design" process or impacted by its wasteful consequences. An example of Toyota's lean PD process is the *Prius* hybrid. At the time this model was being developed, there was intense pressure to meet aggressive time lines set by the company president, Mr. Okuda. The timeline was shortened and shortened despite the fact that the vehicle was to have an entirely new power train (type 2 in the framework laid out earlier with some type 1 characteristics of brand new technology). The simplest solution for the chief engineer, Mr. Uchiyamada, was to short-circuit the process. But Mr. Uchiyamada was a true Toyota man, a chief engineer whose father was also a chief engineer. He refused to compromise, insisting on adherence to the established Toyota process of considering broad alternatives and gradually narrowing the alternatives until a superior engine, body style, and transmission could be selected.

For the body, four design studios (California, Paris, Tokyo, and Toyota City) participated in an open competition, submitting 20 sketches. Of these, five sketches were selected for further consideration and four were eventually turned into life-sized clay models. After broad input from employees, two of these clay models were chosen—one from California and one from Japan. The California design, not surprisingly, was more radical and posed potential manufacturing problems; the design from the Japanese studio was conservative but would be easier to produce. Chief Engineer Uchiyamada asked each studio to try one more time to come up with a truly exceptional design that was also practical. The resulting two models were subjected to extensive feedback from Toyota employees from diverse backgrounds and departments, and their responses were almost equally split between the two alternatives. Reviewing the responses again, Uchiyamada discovered that younger employees and female employees preferred the California design. Based on this distinction, he selected the

California design, which dovetailed with a major Toyota goal: *increase sales among the young and female buyers.*

In a separate, parallel process, the body engineering organization was already working intensely, studying the body and developing solutions. Uchiyamada had told these engineers to assume the design from Japan would win, an assumption probably based on past experience, so they began working on some preliminary structural engineering. After Uchiyamada's decision to go with the California design, they discovered that there were enough similarities between the Japan and California designs that the engineering work they had started was a useful foundation on which to build. The sketches and the intense discussions between engineering and styling in the *kentou* stage had already led to important modifications, joint solutions that satisfied styling, product engineering, and manufacturing. This turn of events attests to the inherent strength of set-based concurrent engineering in the PD process.

Set-based concurrent engineering considers the different design perspectives proposed by different parties, a phenomenon that can best be graphically illustrated by Venn diagrams (see Figure 4-2). Each party has some acceptable range of alternatives—a solution space that will work from its own perspective. Front-loading finds where the sets overlap and, in the process, identifies the winning design solution.

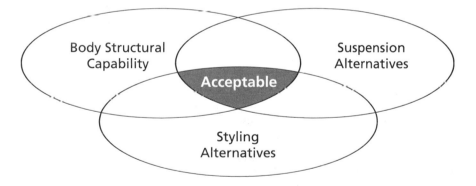

Figure 4-2. Set-Based Perspective—Looking for Intersections between Feasible Regions of Functional Groups

A comparison of Toyota's convergent approach and NAC's iterative approach clearly illustrates the advantages to be gained by the former: Toyota's approach has essentially eliminated a great deal of waste while achieving a superior solution (see Figure 4-3). Where Toyota's deliberate examination of sets of alternatives and systematic convergence is very smooth, NAC's iterative approach has resulted in jerky rework caused by forcing premature decisions based on relatively narrow solution sets.

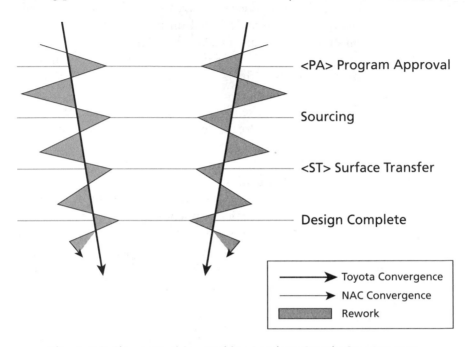

Figure 4-3. Slower Decision Making Leads to Steady Convergence, Forced Premature Decisions Drive Rework

Our discussion of set-based concurrent engineering would not be complete without noting an important underlying premise of the set-based engineering process. Toyota engineers have a strong sense of the vehicle as a system and consequently focus a great deal of skill and energy at the design interfaces. This includes not only the interdependencies of the various individual components that make up the vehicle, but also the downstream manufacturing processes. Consequently, Toyota's process does not focus on the speedy completion of individual component designs in isolation, but instead looks at how individual designs will interact within a system before the design is complete. In other words, they focus on system *compatibility* before individual design *completion*. The principle of compatibility before completion is fundamental to the set-based approach and is a major contributor (along with standard architecture and processes) to Toyota's extremely low number of engineering changes. System compatibility (including design compatibility with lean manufacturing) is a key determinant for attribute selection during the set-based convergence process.

The practice of developing and presenting multiple alternatives is fundamental to both the Toyota PD and lean manufacturing systems and is the basis for decision matrices and other tools discussed later in this book. This chapter provides examples of how each contributor to the development process naturally presents multiple alternatives or solutions to any problem. To get at the root of "set-based thinking," however, requires an understanding that it is more than a specific set of tools or methods. It is, in fact, a cultural characteristic that is woven into "the way things are done." Fortunately, companies can adapt structural aspects of this culture to their own PD by using the following:

- Intentionally identifying multiple solutions to design problems before selecting just one.
- Encouraging engineers (both upstream and downstream) to discuss alternatives early before a fixed decision has been reached on a single design from one perspective.
- Using set-based tools (discussed later in the book) such as trade-off curves to identify the trade-offs of various solutions from different perspectives.
- Capturing past knowledge in checklists (described later) in the form of graphs and equations that show the effects of different alternatives.
- Using system methods like parametric design that quickly show system impacts when parameters are changed.
- Having structured time early in this front-end period for participants with diverse perspectives to work out solutions when the broadest set of alternatives is still available.

Toyota Body and Structures Engineering—*Kentou*

Pre-clay freeze activities mark the beginning of a crucial period during the product development process at Toyota. Referred to as *kentou* or study, this is a time of intense engineering activity throughout the PD program, during which engineers generate hundreds of *kentouzu* or study drawings. At this time, cross-functional Module Development Teams (MDT) meet with each participant to study his or her unique technical perspective of the clay models and design proposals. As noted in Chapter 3, MDTs are cross-functional teams typically organized around the various vehicle subsystems. They consist of representatives of the functional organizations responsible for some aspect of subsystem development, such as the closures or the instrument panel. A typical MDT might comprise one to three senior engineers from the body engineering organization, someone from styling, one or more members from the production engineering group, and, if necessary, an electronics engineer. Pilot team leaders from the assembly plant might also participate if there are important implications for final assembly in a particular subsystem.

This discussion of MDTs requires a brief examination of styling at Toyota, which has intentionally created two different roles for this function. The first is purely artistic, a designer who develops the sketches that may, if selected, become clay models. This designer is relatively protected from discussions of manufacturability and the usual consensus building that characterizes Toyota. The second role is a combination of art and pragmatism, the production designer responsible for modifying a clay model to make it more production ready. This second role is part of the MDT; in fact, some production designers are stationed in the engineering offices and work side by side with body engineers. They represent styling from a dual perspective: one that deeply understands what the customer values and one that deeply understands the permissible degree of freedom to modify a style in order to make it more producible. In organizational terms, a production designer's function fulfills a *liaison role*.

The entire MDT's function is to identify and resolve technical problems and map strategies to achieve component/subsystem-level goals that are aligned with and support overall vehicle objectives. A major part of this task requires studying the design interfaces. As most experienced engineers know, most problems occur at component intersections. It is relatively straightforward to design an outer door. The true challenge lies in designing the door system (inner, outer, reinforcements, electronics, etc.) and ensuring that the door fits the body shell, matches the fender, and the center pillar and to design this system in a manner that supports lean manufacturing. This means making sure that designs are compatible and feasible *before* they are complete. By completing individual designs too quickly, you run a great risk of having to change them later in the program when choices are more limited and expensive, that is, *after* you have matched them up with related components or assessed them for manufacturability.

The importance of using early cross-functional teams for component-level goal setting and problem solving to resolve such challenges cannot be overemphasized. MDT participants are also Toyota's best and seasoned engineers. Using Toyota's standards and fundamental engineering tools and practices, they help identify and reduce variability. An MDT may spend hundreds, even thousands, of engineering hours working toward this goal, so downstream activities can focus on execution.

Standardizing Lower-Level Activities Enables Quick Problem Solving: An Example

Toyota's tradition of standardizing lower level activities enables a problem-solving methodology for finding multiple alternative solutions to engineering challenges. For example, in the design and development of an outer hood panel, manufacturing standards require that the hood be stretched (not drawn because feeding material will reduce required stretch for class-one surface appearance) and that the hood be fully manufactured in three operations (draw, trim, and flange). Body engineering must engineer a hood that matches the required body sweeps at the fenders, windscreen, and grill, while maintaining the vehicle trademark power dome at the center of the hood. The body engineer will create (in section form) several versions of the hood that meet the outlined criteria but will look substantially different from each other. In the meantime, the production engineer assigned to the program (the Simultaneous Engineer) will evaluate these drawings against his:

- Common architecture-standard construction sections
- *senzu* (a manufacturing intent drawing of the previous hood for this vehicle)
- *component quality matrix* (individual part quality plans discussed in Chapter 15)
- the specific goals resulting from competitor tear downs

Using this standardized material and the component quality matrix, the production engineer (PE) then provides written feedback to the body engineer on each proposal. If necessary, the PE may even surface the sections and run them through simulation software, although this does not often occur because Toyota's component-specific process histories are usually sufficient. In any case, the PE can provide quick feedback on the various alternative designs because of standardization and experience.

The PE, based on his significant experience, knows that as long as the periphery depth of the hood is kept below a specific level and is constant, the depth and shape of the power dome and the sweep of the fender lines are of little concern to production engineering. However, these factors do become important near the windscreen surface blends. Here the PE consults previous *senzu* for creating a temporary "manufacturing surface" that is designed to maintain surface tension for good appearance. The PE continues to address specific difficulties and challenges in a systematic, ongoing process, eliminating some alternative designs and combining specific characteristics of others to produce a final design.

This collaborative, simultaneous engineering process has important implications for the PD process and the sense of ownership of key participants. For instance, requiring that team participants closely interact to evaluate multiple solutions against specific criteria is a driving force for generating a breadth of options. And the *kentouzu*, or study drawings, generated from this process are the first step to creating *K4* drawings (to be described later). This process is markedly different from processes practiced by many of Toyota's competitors where, as was noted in the description of NAC, "early feasibility assessment" often means a manufacturing engineer looking at the clay model and providing a list of what seems infeasible. Styling then responds to this "point-based" input with some iteration, which satisfies some concerns and ignores others.

At Toyota, such an approach would be untenable. Instead of styling doing most of their work and then asking for a reaction, the team jointly develops the body with intensive interaction and many data sources at its disposal. The team utilizes competitor teardowns and preprogram plant visits to act on specific objectives defined in the CE's concept paper. The primary tasks of the team during this period are to work together to translate the CE's customer-defined value into meaningful engineering requirements and to resolve subsequent design/manufacturing challenges while staying true to the stylist's artistic rendering of what is a visually appealing design from a customer perspective.

Application of Common Architecture and Principle of Re-use

Another critical aspect of the *kentou* phase is the application of common architecture through detailed design standards and specifications, which the body engineer can draw from a database of best-body sections for each vehicle type. Body engineers can expand, shrink, or otherwise modify these structural best practices while the database simultaneously maintains critical geometric relationships to preserve product performance and manufacturability. Wherever possible, the body engineer identifies carry-over or cross-platform parts for possible use. The principle of *re-use* is crucial to efficiency and quality because qualified and established components substantially reduce performance variability in product development, tooling development, and final production. The downstream impact of re-use on quality and efficiency is dramatic. There are typically two or three body engineering supervisors participating in this stage of the process and holding key meetings with the CE Team and upper management to begin the vehicle-vision cascade.

As the CE narrows the alternative clay models to two or three, digital scans are sent to the body engineering group for more detailed evaluation. This stage of *kentou* is referred to as the *idea proposal stage* and as many as ten core body engineers may be involved. This is also the beginning of *kentouzu* or study drawing generation. Given that the engineers are working from scan data, the majority of study drawings, originally created by hand, are re-created in CAD.

Evaluating and Deciding on Vehicle-Level Goals

The MDT evaluates the impact of vehicle-level goals on the body, identifying and resolving potential problems. For example, the CE Team may have decided to incorporate a new lighting system that was demonstrated by one of Toyota's suppliers, and this may have a significant effect on the design of the front fascia and fender and, in turn, on the manufacture of those components. The body engineer generates several solutions to this design challenge for the production engineer to evaluate quality and manufacturability. It may be that the shape of the new lighting system, when combined with the fender-to-hood line sweep, will cause the "nose" of the fender to be too sharp and thereby create a poor quality manufacturing condition. Or perhaps a new safety regulation in the United States requires increased bumper impact absorption, which, in turn, will require greater bumper overhang that affects vehicle styling. These kinds of challenges will also initiate multiple alternatives for the MDT to review. A great deal of negotiation takes place in these teams, and participant passion can lead to conflict. However, Toyota's "customer first" mentality or customer-defined value, is the final arbiter for decisions.

During this time, there is also a great deal of component or subsystem testing. Wherever a nonstandard condition exists, the body engineers carry out virtual or physical testing of rapid prototypes and mockups. Although testing is often straightforward, it is always scientific.

Body engineering holds regular (often weekly) meetings with several module teams (or the "cross-module team") to address detailed technical problems and decide schedules, budgets, and design interfaces. These meetings are typically short because the engineers use a variety of communication tools (described in Chapter 14) and hold small pre-meetings (often consisting of two engineers) to provide all relevant problem descriptions, Toyota standard references, and proposed solution alternatives to the appropriate meeting participants before the meeting. Participants are expected to be informed and prepared for the meeting and to have achieved consensus. At the meeting, team members engage in a detailed and productive discussion. This type of lean cultural behavior is another illustration of integrating process, people, and tools and technology.

As the vehicle design matures and the clay models are reduced from two to one, body engineers begin to transform their study drawings into simultaneous engineering drawings to initiate several specific production engineering activities, such as processing, fixture layout, or binder development. The design intent of these drawings is close to the final layered drawings that will be available at final data release. Given the standardized processes and the size and general shape of parts, the product engineers can begin their simultaneous activities in earnest.

Toyota Production Engineering: The Simultaneous Engineer's Responsibilities

In the 1990s, Toyota continued to push manufacturing considerations further up in the development process. Body engineers had always been very cognizant of manufacturing issues and had spent time working in the plant as well as regularly visiting the plant, but Toyota wanted more. Under its current, more aggressive simultaneous engineering process (in the early stages of *kentou*), key production engineers are assigned to vehicle program MDTs as simultaneous engineers (SEs) who function as full-time representatives of their manufacturing discipline. For example, the body closures MDT might have an SE lead and several SE members who are each responsible for 12 or more individual parts throughout the entire program.

At the beginning of the *kentou* phase, the SE team studies electronically transmitted design information before attending the biweekly MDT meetings. Body engineering hosts these meetings and with production data and updated *senzu* (manufacturing drawings) and checklists in hand, the engineers discuss the analysis of current design proposals and participate in the often arduous negotiations. They will spend hundreds of hours analyzing, discussing, and codeveloping multiple-design alternatives to achieve process and product goals.

This is an intense period for each simultaneous engineer, during which single MDT sessions may run for several 12-hour days at a stretch. Since the SE is responsible for his or her parts from this time until start of production (SOP), there is strong motivation to evaluate the quality and productivity implications for manufacturing of the proposed designs thoroughly. The SE will serve as "Lead Manufacturing Engineer" and will be fully responsible for a specific set of parts as he or she shepherds them through the entire development process. This process eliminates many of the handoffs and associated wastes in a typical product development process. The SE is also aware that the decisions reached now will affect the success of every other step in the development process to SOP.

As noted above, Toyota uses the CE's concept paper, field QA data, competitor teardown sheets, and current production process data to establish specific goals for both the vehicle system and individual components. Toyota accomplishes this by translating specific product quality and performance goals, based on customer-defined value, into specific part design characteristics. This, in turn, requires a robust manufacturing process capable of consistently delivering parts within an acceptable tolerance range at a reasonable total cost. Each body engineer and simultaneous engineer understands that this process will require saying no and eliminating non-critical part requirements or unreasonable tolerance bands. The process also identifies critical customer-defined quality characteristics, and eliminates nonessential elements in the pursuit of those specific goals.

SEs Must Hit Investment and Variable Cost Targets

The SEs are also responsible for hitting both investment and variable cost targets for their parts—both the tooling cost and the parts produced by these tools. This is the beginning of enabling lean manufacturing. It is the CE who sets goals for performance and quality for both the part and the efficient manufacture of the part. It is also the CE who establishes the cost targets with a continuous improvement concept in mind. It is the role of the SEs to work early and hard in the PD process to meet and design (in process) capabilities and efficiencies for both the product and process. This is a far cry from teams that try to "thrift" or improve a process after the fact.

Lean manufacturing processes are standardized by part, continuously improved, and adhered to by all participants. The key to SEs successfully managing such broad responsibilities and achieving cost, quality, and performance targets is to front-load problem solving and adhere to disciplined standardization. This makes the body engineer and the simultaneous engineer equally responsible for adhering to standardized manufacturing processes. The early design process consists of the following:

- defining a "design space" or system requirements
- creating multiple design and process alternatives (or solutions) based on standards (including common construction sections)
- quick testing and program objectives, analyzing each alternative's impact on cost, quality, and performance
- rigorously honing in on the essential characteristics of each alternative
- combining characteristics across alternatives
- focusing energy and effort toward a single design and process solution

In this sense, product and process are "codeveloped" and solutions are "designed in" instead of becoming expensive add-ons later in the process.

Mizen Boushi and Going to Production Plants

In preparation for the *kentou* process, the SE spends a good amount of time in the production plants gathering data and talking to team leaders and operators in order to understand fully current manufacturing issues and solicit potential countermeasures. If necessary, the SE invites production team members to visit with the MDT. This is an important part of the *mizen boushi* or designed-in quality process during which engineers focus on "designing in" countermeasures. Roughly translated, *mizen boushi* means "to prevent mistake" or "preventive measure" and refers to a disciplined process that focuses on the early design stage to engineer products and processes that support lean manufacturing and produce robust quality. Standardized design and process checklists (discussed in Chapter 6) are crucial to this very rigorous process.

Communicating with Functional Specialists

During the *kentou* process the SE continually communicates with different functional specialists within the production engineering department, such as process engineers, die designers, fixture and assembly engineers, and even with assembly teams in the plant. Although he or she is an experienced production engineer, the SE is not typically as knowledgeable as the specialists and utilizes their specific knowledge to deal with highly technical or unusual situations. In this way, the SE acts as the link between styling, body engineering, production engineering, and the production plant. As such, the SE communicates on both technical and logistical issues, helping to keep the activities of both groups synchronized. From the beginning, the SE communicates specific performance, cost, and quality objectives with the functional specialists and uses their input to fashion his or her counterproposals for the MDT. In addition to obtaining valuable technical input, this activity also creates early familiarity and a sense of ownership among the functional specialists on whom the SE will rely for much of the core manufacturing engineering tasks. This is crucial to achieving both PD and manufacturing process efficiency goals. The SE also works with specific suppliers if these provide manufacturing engineering services. This activity in and of itself can enhance the PD process.

The SE Submits the Plan

The SE eventually emerges from the communication part of the process with a deep understanding of the product goals and a process plan. The SE prepares, on an 8.5 × 11 sheet of paper, a process plan for each of his or her components and subsystem, which is reviewed with the appropriate functional specialist within the production engineering organization. This plan, along with the part-specific *senzu* and quality matrices, is used to begin preparation for such activities as "prepare for die/fixture design" and "preliminary binder developments" in the core production engineering groups.

Leveraging Digital Tools

Much of Toyota's ability to front-load its PD programs has been enhanced by significant advances in digital technology (discussed further in Chapter 13). PD teams can begin very early in the process with advanced design tools such as the CATIA CAD system, enabling design fit-up analysis or "digital mockups" that identify design crashes (interferences) and ratholes (voids). Often, the process includes parametric design models to make certain that any change in part design is matched by updating all related parts and tools. Sophisticated CAE crash and manufacturing simulation tools enable short problem-solving cycles and allow more iterations to run earlier, in less time, and at a lower cost. In fact, in many cases, these tools have eliminated the need for elaborate, expensive, time-consuming physical prototypes. These technological innovations continue to shorten lead-time, reduce cost, and improve product quality.

Early Problem Solving in *Kentou*: A Case Example

The lean PD process consists of an interrelated series of problems that must be solved: technical, logistical, and financial. Obviously, organizations with excellent problem-solving skills will achieve higher quality solutions faster and have a significant competitive advantage in product development. In addition, companies that are superior at problem solving will learn from their experiences and have a greater knowledge base from which to draw. This means they will spend less time "re-solving" problems.

Kentou results in far fewer engineering changes and creates process flow by allowing companies to focus on downstream task execution. *Kentou* also provides a formal structure for cross-functional teams to "design in" solutions. This, of course, is far less expensive than solving problems or "fixing" designs later in the process.

One fairly straightforward example of this problem-solving characteristics of *Kentou* occurred during a *Camry* development program. The new style called for a wrap-around headlight design that would have design, engineering, and manufacturing implications for the fender, hood, and front fascia. As the MDT worked its way through these challenges during the *kentou* phase, a critical problem surfaced. The new headlight clearance requirements were driving a fender design with a long, narrow nose (see Figure 4-4).

The production engineering group suspected that this condition would be problematic both in terms of dimensional stability of secondary forming operations (twisting caused by material springback) as well as for material handling (the narrow condition on the nose would be easily damaged). Sketches and digital simulation tools confirmed these suspicions and allowed the MDT to design and confirm a solution (see Figure 4-5). By making a minor modification to the hood cut line (edge) the team was

Design Issue

New wrap-around type headlight system required on vehicle, which will necessitate a larger clearance area on the fender.

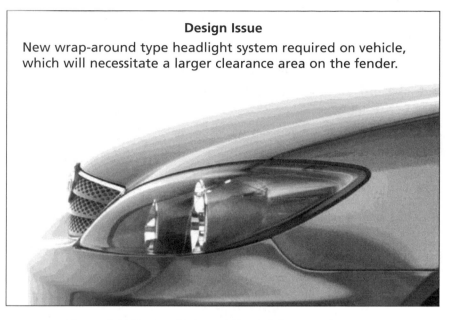

Figure 4-4. *Kentou* Early Problem Solving—Design Issue

Analysis

Early sketches showed a clearance opening that required a long narrow nose condition on the fender.

Narrow Nose
Condition

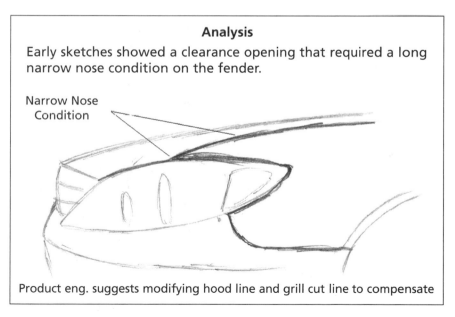

Product eng. suggests modifying hood line and grill cut line to compensate

Figure 4-5. *Kentou* Early Problem Solving—Analysis

able to shorten the nose of the fender and arrive at a condition that satisfied all participants (see Figure 4-6).

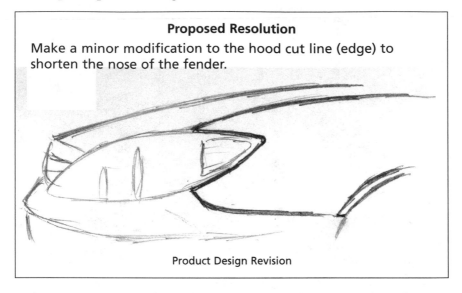

Proposed Resolution

Make a minor modification to the hood cut line (edge) to shorten the nose of the fender.

Product Design Revision

Figure 4-6. *Kentou* Early Problem Solving Example—Proposed Resolution

Had the team discovered this problem later in the process, there might not have been sufficient design flexibility to allow for a designed-in solution. This would have led to a compromise based on cost of the engineering change—an inferior solution.

An example of "early problem solving" at the North America Car Company (NAC) illustrates the stark contrast between a traditional and lean PD approach. NAC recognizes the value of solving problems earlier in the process, but has not developed any mechanisms such as MDTs or a *kentou* period for this purpose. During the prototype phase of a recent program, a supplier noted that the design condition of the inner lift gate was very similar to the design of the inner lift gate from a recent program, which resulted in material compression and wrinkles on the seal surface (see Figure 4-7).

Unfortunately, the design solution required to rectify this condition would change other adjacent parts that were also in the prototype stage (see Figure 4-8).

Since these designs were "mature" and tooling for these parts had already started, the only possible fix was to change the tools for manufacturing

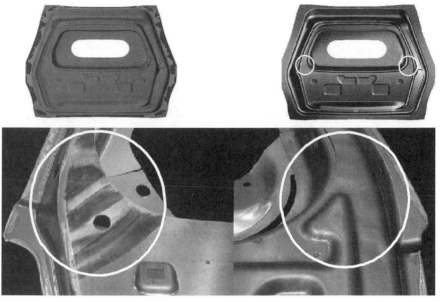

Current Program   **Previous Program**

Figure 4-7. Seal Surface Compression and Wrinkle Condition

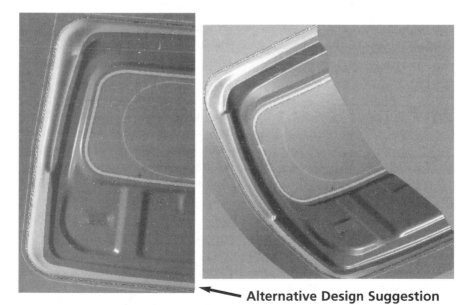

← **Alternative Design Suggestion**

Figure 4-8. Alternative Design Solutions to Seal Surface Condition

the inner lift gate itself, which would help to minimize, but would not eliminate, the wrinkling condition. Changing the massive stamping tools was not only expensive, but also required considerable time.

Kozokeikaku (K4) Pulling the Pieces Together

The *kozokeikaku* or K4 is a high-level body structures document that pulls together the final set of individual *kentouzu* into a body system plan. The K4 term has evolved among Toyota's North American Engineers because of the difficulty of pronouncing *kozokeikaku* and the four k's contained in the word. The K4 plan includes critical vehicle cross-sections, locators, clearances, and design intersections. It will also call out specific assembly requirements, critical tolerances, and any other important manufacturing direction as well as any potential standard process deviations. The K4 is the body system execution plan. It defines all vehicle system requirements for individual designs, provides crucial guidance to the detail design process and is an invaluable aid in the real world execution of *compatibility before completion*. It pulls together the results of the many study drawings developed in the *kentou* period. The K4 is circulated to all concerned functional groups in both PD and Manufacturing for signatory approval. The K4 is developed toward the later part of *Kentou* and is typically released about a month after styling completion and provides the basis for early system analysis, manufacturing planning, and all subsequent detailed drawings.

Right Person, Right Work, Right Time

Unfortunately, the early stages of product development at many companies are poorly understood and unstructured. As a result, they get few resources and little attention. In these companies, management is unwilling to commit its most capable human resources to the early stages of a program, choosing instead to assign its best people to fire fighting in the final stages. This is a waste of both corporate resources and the potential and morale of high-performance individuals. Product development is talent driven, and talented people generally prefer creating something great rather than patching up a sinking program. Companies that run ineffective PD programs often suffer from *brain drain* and constantly produce mediocrity.

Getting the most out of talent is about using process, people, and tools and technology effectively. Companies that do not get serious about product development until they begin to review completed designs, or worse, physical prototypes, continually struggle with massive cost over-runs and delay-riddled product introductions. Part of the problem is that they subscribe to the notion of a "fuzzy front-end" and a "fix things" later philosophy. When the front-end of concept development is "fuzzy," it suggests that serious, disciplined processes are impossible in this early stage. Lean thinking sees this undisciplined approach as a slow, expensive, and unpredictable process that can seriously disrupt product develop-ment and other activities.

The concepts of concurrent engineering and front-loading have been around for a while. As a result, some companies, in their eagerness to speed up their PD process, have attempted to push more work up front without a detailed understanding of the implications of this decision. This is "front-load by brute force" and invariably leads to mistakes (such as insisting on final pricing from suppliers before designs are complete or doing too much process work with immature designs or rushing to com-plete individual designs earlier in the process). These, and similar mistakes lead to greater and greater levels of rework and a slower overall process.

Toyota, in contrast, focuses on doing the right work at the right time executed by the right person. This in part explains the findings of Ward, Liker, Sobek, and Cristiano (1995) that delaying decisions at Toyota gener-ally leads to faster overall product development.

In the next chapter, the discussion returns to minimizing the second PD waste, product development process waste. The chapter also looks at useful tools for combating this waste.

LPDS Basics for Principle Two

Front-load the process to explore alternatives thoroughly

The LPDS is front-loaded because the start of the program is where you can have the greatest impact on the success of the product for the lowest cost. Preprogram, multiprogram management activities include product portfolio management, technology and platform planning, PD system resource management, and the management of shared program content to create an environment where individ-ual programs have the best chance of being successful. In individual programs, cross-functional teams made up of the most experienced people come together at the start of the project to look at a broad range of solution sets that anticipate and solve problems, to design in countermeasures for quality and manufacturability, and to isolate the inherent variability in product development in order to facilitate flawless execution in the next phase of the program. At Toyota, this takes place during an intense, initial program phase referred to as *Kentou*, during which hundreds of *kentouzu* are generated and Toyota's most experienced engineers resolve approximately 80 percent of a program's technical problems and dramatically reduce late engineering changes.

CHAPTER 5

Create a Leveled Product Development Process Flow

At Toyota we try to make every process like a tightly linked chain—where the processes are connected by information and by physical flow. There's nowhere for a problem to hide. The chain never works perfectly. But if we know where our breaks are and our people are trained to fix the breaks, we get stronger every day in the company. It keeps us on our toes, it self-identifies muda *and, five whys is our method to eliminate* muda.

GLENN UMINGER, Toyota Manufacturing Corporation, North America

The Power of Flow

In 1913, Henry Ford and his team revolutionized the world of manufacturing by developing a continuous assembly line for the Model T Ford and introducing the world to the power of flow. Ford's dramatic results were not lost on Toyota, which was quick to learn from Ford and has subsequently become the leader in adapting the concept of flow to a variety of environments. For Ford, however, flow stopped once a process moved away from the assembly line. Outside the assembly line, large batches of parts made by machining operations, stamping operations, and injection molding operations were pushed, rather than pulled, to the assembly line.

Using cellular manufacturing, Toyota extended the concept of "one-piece flow," or leveled flow, throughout its operations—even into supplier operations. When it was not possible to flow one piece at a time, Toyota built small stores of parts, sometimes called "supermarkets." This supermarket concept reflects what actually occurs in real supermarkets: Customers "pull" what they need from store shelves and the owner or manager replenishes the shelves as needed, restocking what customers have purchased. In manufacturing industries, activity provides a direct link between the customer and the supplier of materials in a pull system. The smaller the batch size the producer makes, the closer the operation is to the ideal "one-piece flow." To maintain leveled flow, producers must be wary of waste. Toyota has long been aware of this and conducts its "war on waste" on processes companywide.

So what does this have to do with product development? This chapter posits that Toyota's success in product development starts with viewing PD as a process. Like any process, PD has a cadence and repeated cycles of activity. Toyota has done an exceptional job of standardizing the PD process to bring to the surface the repeated cadence that allows continuous improvement through repeated cycles of waste reduction. The chapter focuses on the notorious seven wastes of lean manufacturing and identifies some of the solutions Toyota has discovered to remove these wastes from the product-development value stream. In addition, this chapter shows that Toyota has managed to "level the flow," not only by eliminating waste (*muda*) but also by eliminating "unevenness" (*mura*) and "overburden" (*muri*). In this way, Toyota has steadily moved ahead of its competition in bringing products to market more quickly and with higher quality.

Viewing Product Development as a Process

As discussed in Chapter 4, the majority of Toyota's product development is *derivative product vehicles built on existing product platforms*, focusing on modifications of existing products or variations on a theme. This type of PD system typically has a great deal in common with other process-based systems, such as manufacturing operations. For instance, modern PD systems must deal with multiple projects simultaneously and face similar shared resource management challenges. Moreover, even though many of the specific design challenges might be different, the basic work, the tasks, and the sequence of tasks, are the same across programs. From this perspective, companies can view the PD system as a *knowledge work job shop* that must deal with multiple work centers, constraints, and an integrated network of queues. This PD process perspective enhances a company's ability to apply adapted process management tools and methods to reduce variation and create leveled process flow without destroying the creativity necessary for great products.

This view of product development begs the question of just how "lean" applies to a process in which information flows is more important than material flows and in which variability from project to project is a given. In other words, can PD value streams be made to flow as smoothly as the manufacturing of physical goods? There are two answers to this question: yes and no. Yes, you can view product development as a repeatable process of steps that are interrupted by waste. Stabilizing the PD process and improving it through waste reduction is quite possible and effective. But no, it is clearly not identical to repetitive manufacturing processes. The tasks are more complex with relatively long cycle times and far too much variability to pull out seconds. Thus, even though many of the concepts and methods discussed in this chapter have common roots in manufacturing—as discussed in the *The Toyota Way* (Liker 2002)—it is critical to keep in mind that product development has its own complex environment and fundamentally unique challenges. One of the first of these is *seeing the process*.

Value Stream Mapping

The key tool for understanding the flow of material and information and seeing the manufacturing processes is value stream mapping (Rother and Shook, 1998). This methodology looks at the transformation of material as a series of process steps interrupted by waste. What governs the flow is information that tells individual processes what to make, how much, and when. By mapping the current state and identifying the waste that inter-rupts flow, value stream mapping then moves to a leaner future state vision that is translated to an action plan. This powerful tool seems to be fine for repetitive manufacturing processes but is, at least in its original form, difficult to apply to product development.

With modifications, however, value stream mapping can become a powerful tool for improving product development value streams. Morgan (2000) has successfully adapted value stream mapping to the complex PD environment. (This modification, called PDVSM, is reviewed in Chapter 17 and illustrated in the appendix to this book.) The starting point for representing the PD value stream is recognizing it as a process. One key attribute of this unique process is that it involves many parallel, interde-pendent activities rather than the neat serial value streams typical of manufacturing.

In any given product development project, there is a lot going on. The primary activities are value-adding activities and waste is hidden. The key to superior product development is to disentangle this complex web of activity into definable "work streams" that perform a distinctive function: converting inputs into outputs. Doing this reveals a number of parallel work streams that are, to some extent, independent but are also inter-dependent. It also reveals waste within each of the work streams.

Figure 5-1 illustrates a PD value stream as a series of parallel "work streams." Each work stream (or "swim lane") has a set of serial processes interrupted by waste. The figure shows the work streams on a common timeline; viewing a single slice or section of this reveals what is happening across work streams. As Chapter 17 of this work will show, the PDVSM tool can also be used to reveal connections across work streams, different types of waste, and even feedback loops.

One thing that will be apparent to any one experienced with product development is that the timing suggested by a chart like this is rough at best. For example, not every project will have a concept stage that starts or ends as neatly as this diagram suggests. It is not always clear exactly when a concept stage begins or ends. For this reason, it is important to note that this diagram illustrates only concepts. Actual current-state value stream mapping should reflect the activities and timing of a real product development project—not an abstract "typical project." All projects are somewhat unique. At this level, the task at hand is to understand the wastes and sources of waste in a product development value stream, not to measure precise and invariable activity sequences and timing.

With this caveat, it is important to reemphasize that the starting point for improving a process is to understand it as a process. And the starting point for eliminating waste is to recognize waste. Figure 5-1 suggests that (from the perspective of the information being transformed into a design) most of the time spent in a product development process is, in fact, waste. But what most companies typically do in the name of reducing PD lead times is to focus on reducing time in value-adding processes. For example, they implement a new CAD feature that reduces the time for entering data into the computer and creating the 3-D graphic. Or they buy faster computers to make the engineering analysis algorithm converge more quickly. Lean, on the other hand, begins with looking at the total value stream because waste between process steps is likely to be far greater than waste within a single process step (like entering the CAD data). A full appreciation of the concept of waste requires characterizing waste in PD value streams and understanding its causes.

Seven Wastes in the Product Development Process

Most product development processes are not lean. They are, in fact, fraught with waste. Waste or *muda* is any activity in a process that consumes resources without adding value for the customer. In this work, the

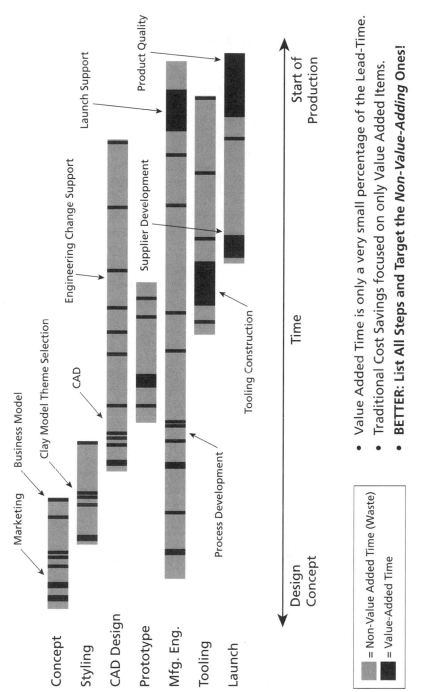

- Value Added Time is only a very small percentage of the Lead-Time.
- Traditional Cost Savings focused on only Value Added Items.
- **BETTER: List All Steps and Target the *Non-Value-Adding* Ones!**

Figure 5-1. Product Development Lead Time

starting point for identifying product development wastes is Taiichi Ohno's seven categories of waste: Overproduction, Waiting, Conveyance, Processing, Inventory, Motion, and Correction. Although causes for these notorious seven are different for manufacturing, these same categories can be quite useful in revealing non-value-adding activities in product development (see Figure 5-2).

Seven Wastes	What is it?	PD Examples
Overproducing	Producing more or earlier than the next process needs	Batching, unsynchronized concurrent tasks
Waiting	Waiting for materials, information, or decisions	Waiting for decisions, information distribution
Conveyance	Moving material or information from place to place	Hand-offs/excessive information distribution
Processing	Doing unnecessary processing on a task or an unnecessary task	Stop-and-go tasks, redundant tasks, reinvention, process variation—lack of standardization
Inventory	A build up of material or information that is not being used	Batching, system overutilization, arrival variation
Motion	Excess motion or activity during task execution	Long travel distances/ redundant meetings/ superficial reviews
Correction	Inspection to catch quality problems or fixing an error already made	External quality enforcement, correction and rework

Figure 5-2. Applying the Seven Wastes to Product Development

1. *Overproduction:* In manufacturing, this is producing ahead of what is actually needed by the next process or customer. In prod-

uct development, this waste is common where processes are not well synchronized across functional organizations. Examples of this include any task that is completed before the next operation is ready to process it or conversely, downstream operations working on upstream designs prematurely in an effort to do concurrent engineering. Another example of overproduction is working on the wrong activities instead of on activities the next process really needs. Often, overproduction is the result of completing design work before checking for its system compatibility or manufacturability.

2. *Waiting:* In manufacturing, operators stand idle waiting for material or do nothing while their automatic machines are processing. In product development, engineers seem to be in a state of perpetual motion, always rushing from meeting to meeting or engrossed in something on a computer screen. From the perspective of a workstream, however, there is often some key activity that engineers should be working but cannot because they do not have what they need to proceed with the given task. Before they can proceed, engineers routinely wait for reviews, decisions, permission, information, purchase orders, or some other wasteful transaction activity. Our experience has repeatedly demonstrated that waiting is one of the most pervasive wastes in the PD process.

3. *Conveyance:* In manufacturing, this means moving parts and products unnecessarily. In product development, it means unnecessary handoffs from one overly specialized activity to another, and more specifically, information changing hands, whether by word, picture, or data exchange. This waste leads to the loss of momentum, information, and accountability in the process. It is also a commonly accepted dysfunction in traditional PD systems.

4. *Processing:* In manufacturing, this waste is manifested by unnecessary or incorrect processing. In product development, it includes engineering errors or system flaws. Proper training and development can eliminate or greatly reduce the former. It can also result from designing new components instead of using carry-over, designing from scratch instead of morphing standard design architecture, or creating new manufacturing processes for each program instead of working to a standard manufacturing process. Another example of processing waste is unnecessary transactions and negotiations that transpire while selecting and managing suppliers.

5. *Inventory:* In both manufacturing and PD, inventory waste is the direct result of overproduction. For manufacturing, this is simply having more than the minimum stock required for a precisely controlled pull system. In product development, it is excess information, such as designs that wait for the next available resource. Information waiting in queues to be processed represents the most insidious waste in product development. Often, there are problems in this information (e.g., it is not what the next process needs) or information gets lost and is late in getting to where it is needed. These problems often remain hidden; by the time anyone discovers them, they have already resulted in extensive rework and long lead times.

6. *Motion:* In manufacturing, operators move in ways that are unnecessary or cause strain. In product development, engineers attend unnecessary meetings, create redundant status reports, and prepare for and participate in unnecessary project reviews. Motion waste also includes nonsubstantive, "wide aisle" type plant and facility tours that do not lead to the direct data needed to make real design decisions.

7. *Correction:* In manufacturing, this is inspection, rework, and scrap. In product development, it takes the form of program audits, reviews, testing new components instead of reusing proven ones, late engineering changes, excess tool tryout, and all forms of rework. Most product development processes are so inflated with the correction waste that at least a third of total allocated resources are engaged in rework at any given time.

There Are Really Three Ms

People often define the goal of lean as waste elimination. But battling *muda* does not accurately represent all that "lean" is about. True lean thinking does not focus on one-dimensional *muda* elimination; it works to eliminate three types of interrelated waste: *muda, mura,* and *muri*—collectively known as the three Ms.

1. *Muda* (non-value-added). The best known "M," it includes the seven wastes of the Toyota Production System and their lean product development counterparts discussed in Chapter 3. Any activities that lengthen lead times and add an extra cost to the product, for which the customer is unwilling to pay, are considered *muda.*

2. *Muri* (overburden). In some respects, *muri* is the opposite of *muda*. *Muri* is pushing a machine, process, or person beyond natural limits. Overburdening people can lead to sloppy work resulting in quality problems and potential safety risks. Overburdening equipment causes breakdowns and defects. Overburdening a process means long queues that increase PD lead-time or short-circuiting the process, which leads to downstream errors and rework.

3. *Mura* (unevenness). In normal production systems, the workflow is uneven. Sometimes, there is more work than people or machines can handle; at other times, there is not enough work. Engineers are all too familiar with the frenetic pace of work immediately preceding a deadline (for example, prototype review or new product launch), which is generally followed by calm with relatively little work pressure. Unevenness results from an irregular production schedule or fluctuating production volumes caused by internal problems such as computer downtime or missing information. *Muda* will be a result of *mura*. With uneven production levels, it will always be necessary to have the equipment, materials, and people on hand for the highest level of production—even if the average requirements are much lower.

When they begin to apply lean methods, most companies look at any process where the work schedule swings wildly and begin engaging in a "war on waste" to eliminate *muda*. Almost invariably, they start by pulling out inventory and linking operations. They also look at the work balance and take people out of a system; they also reorganize the workplace to eliminate wasted motion. Then they step back and let the system run. To their dismay, the "improved" system often runs itself into the ground! People become overburdened, sick leave soars, equipment break downs more often, and before long, management is convinced that lean doesn't work. What these companies fail to realize in "implementing lean" is that they have done nothing to stabilize the system and create the "evenness" that allows lean tools to work properly.

Lean thinking has made it is easy to identify waste and pull it out of a system, but it takes much more effort to create an evenly balanced lean flow of work. What many companies are fixating on is blindly pulling out *muda* because this can lead to short-term cost reductions. An extreme example of this is downsizing the engineering work size by 10 or 15 percent. Costs go down and the company has eliminated "waste." But has it really? In lean thinking, the real and more difficult challenge is the long-term task of continuously driving out *muri* and *mura—managing and correcting an overburdened and uneven system.* Downsizing engineering work does neither. In fact, it inevitably results in overburdening and unevenness, albeit somewhere else in the process.

Most people assume that continuous flow means *leveled* continuous flow. But the result of this assumption is flawed decisions that eliminate "waste" in one activity while ignoring waste elsewhere in the process. For example, having material flow one piece at a time across work centers without inventory does not correct a jerky and unstable pace and mix of production. This is not and will never be a continuous flow. It is simply an erratic, one-piece flow, with starts and stops, overutilization and underutilization that is inherently failure prone and will almost always fail to create quality, productivity, or continuous improvement. Continuous flow is eliminating *all* non-value-adding activities of a product's progression along its value stream so that it flows *unimpeded continuously*—from concept to delivery.

Barriers and Facilitators of Flow: Insights from Queuing Theory

It is important at this juncture to examine why the seven wastes discussed above are so prevalent because no company can truly understand how to eliminate these wastes until it understands their true root cause. Viewing the PD process as a knowledge work job shop and considering the well-known body of knowledge on queuing theory will provide some important insights about the root cause of waste in product development. Queueing theory can help us to see how traditional approaches to product development actually exacerbate the inherent variability in the process andcause incredible amounts of waste. Traditional PD practices that are particularly problematic include:

- PD work centers working in large batches created by the stage-gate or milestone based product development processes.
- PD work centers with differing levels of capacity at any point in time creating capacity mismatches and a general ignorance of work center capacity and subsequent constant system overburdening.
- Unpredictable PD workloads expanding to take up all the time of all engineers assigned to projects.

- Highly cyclic PD workloads characterized by lulls in workload followed by tremendous system congestion, expanding lead times beyond planned deadlines.
- Low levels of task execution and scheduling discipline, leading to high levels of both task and interarrival variability.

To gain insight into the reasons wastes are generated by this approach consider the PD process as a system—where arrivals (requests for work) place demands upon a finite capacity resource—and apply queuing theory—the phenomena of standing, waiting, and serving. Then consider traditional product development practices in the light of the following basic tenants of queuing theory which are well understood in manufacturing and well documented in *Factory Physics* (Spears and Hopp, 1996):

- *Law of Batches*: "Cycle times over a routing are roughly proportional to the (move) batch size used in the routing."
- *Law of Variability Placement*: "Variability early in the routing has a greater impact on WIP and cycle times than equivalent variability later in the routing."
- *Law of Utilization*: "If a system increases utilization without making any other changes, average cycle time will increase in a highly nonlinear fashion."
- *Law of Variability*: "In steady state, increasing variability always increases average cycle times and WIP (work in process) levels."

These principles can teach us a great deal about the root cause of much of the fundamental waste in the product development system. The "law of batches" suggests that if we manage the product development process as a series of gateways at which the process must pause and then organize the product development system into a lot of separated work centers (sometimes called functions or chimneys), each processing and releasing work in large batches of information we begin to understand why we experience tremendous levels of WIP and incredibly long lead-times in product development. Consider, for example, a centralized engineering analysis department getting many requests for analysis over time and then working on these requests and issuing results in batch mode. The natural result is that this will extend lead times. Substitute for engineering analysis executive reviews, the design studio, body engineering, the prototype shop, testing, CAD designers, tooling designers, etc. and you will get the same results that the law of batches predicts for a job shop.

Capacity over-utilization has well understood consequences for system performance yet system capacity is seldom even a consideration in product development planning. The queuing curve illustrates the increases in lead time based on changes in system capacity utilization (see Figure 5-3). It begins to increase in a nearly exponential manner at about 80 percent of capacity utilization, meaning that there is a nonlinear relationship between additional system loading and lead-time increases when 80 percent of system capacity has been reached. Product development systems often operate at levels significantly above this level.

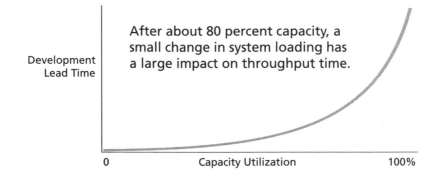

Nonlinear relationship between capacity utilization and lead time causes systems to bog down before full utilization (Factory Physics, 1996)

Figure 5-3. Effect of Overburdening Capacity on Development Lead Time

To make matters worse, the high level of variability naturally associated with the PD process further exacerbates these system-loading effects as illustrated in Figure 5-4

Variability is a primary determinant of poor system performance. Unfortunately, most traditional PD systems are ripe with variability. There are two types of variability to be concerned with:

1. *Task Variability:* This refers to differences in the methods and duration of specific tasks rampant in most product development.
2. *Interarrival Variation:* This refers to the time difference between when work is scheduled to arrive at a workstation and when it

actually arrives. This is often caused by the first type of variation as well as by capacity constraints.

When these types of variations combine within a system, we get a rapid propagation of variation that is incredibly destructive to system performance. Moreover, if the variability is present early in the process, it will have a magnified effect throughout the process. Thus, any efforts to eliminate this variability on the front end (the concept stage as covered in LPDS Principle 2) will pay huge dividends. Consequently, managing variation and controlling system capacity is crucial to high levels of product development system performance.

It may be easiest to understand these phenomena through a commonly cited analogy. At one time or another, we have all been in a traffic jam. Traffic comes to a complete stop and then moves at a snail's pace for miles. Finally, we discover the cause of the delay. Two cars, involved in a minor accident, are on the side of the road with a police car partially blocking traffic in one lane of a three-lane highway (see Figure 5-5). But why would this cause traffic to stop. The fact of the matter is that if the conditions were different (e.g., fewer cars on the road or low highway system utilization), then the partially blocked single lane would not create a jam. Excess highway system capacity would be available to absorb the variability. However, if the highway is 80 percent utilized (or more), losing one

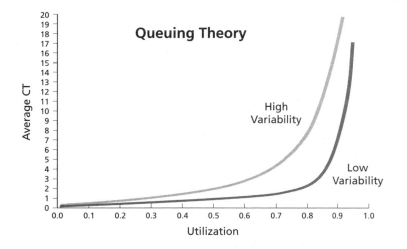

Figure 5-4. High Levels of Variability Further Exacerbates the Effects of Capacity Utilization (*Factory Physics*, 1996)

lane causes a major backup. In this case, the variability in the system (the cars unexpectedly getting into an accident and heavy traffic at that particular time) will lead to total system congestion and long lead times (i.e., traffic moving at a snail's pace).

Low Resource Utilization   **Above 80 Percent Capacity**

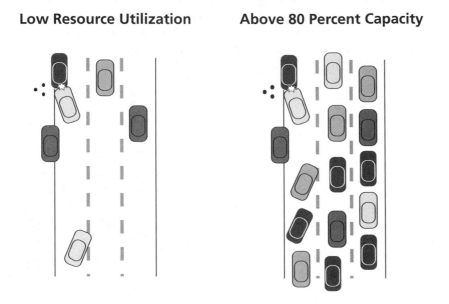

Figure 5-5. Queuing Theory and Traffic Jams

Striking evidence for the effectiveness of applying the principles of queuing theory in product development is provided by Paul Adler. Adler et al. (1996) studied several PD projects and verified a common finding: that the projects progressed fairly smoothly during periods of moderate workloads when people were not very highly utilized. However, when the workload increased and reached about 70 to 80 percent of system capacity, any additional workload began to cause the lead time to increase dramatically. Any variability in the process (e.g., missing information, wrong information, late arrival of key data) interacted with any increases in workload, causing the system to blow up. In fact, overutilization of PD resources, along with high levels of process variability is the primary reason for long lead times, program delays, and even quality issues. By managing system utilization and employing basic tools to reduce both task and interarrival variability these companies experienced both decreases in lead time and increases in overall system productivity over time (Adler et al, 1996).

Queuing theory principles nearly always reveal the root causes underlying the seven wastes of lean product development. Figure 5-6 shows the relationships between these causes and the seven wastes. The figure by no means presents a comprehensive list of all causes; nor does it identify all of the relationships between causes and waste. It does, however, provide a graphic illustration that provides a general insight into the system causes that must be changed by fundamental changes in the PD system.

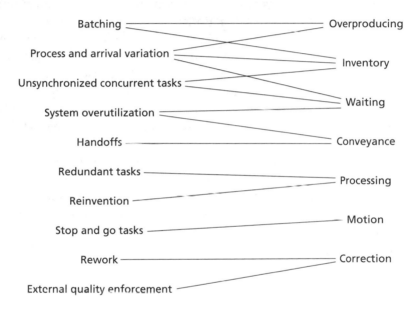

Figure 5-6. System Causes of Seven Product Development Wastes

Companies that continue to view the PD process in the traditional way; as an uncontrollable series of discrete events that just pop up, will always be on the defensive and be forced to deal with variability and capacity issues in a fire drill mode. Toyota never accepts variability problems as necessary or uncontrollable. The following discussion of Toyota's PD process presents the powerful antidotes developed by Toyota to attack waste at the root and create PD process flow. Detailed examination will show that many of Toyota's practices align extremely well with queuing theory fundamentals and that they start from the beginning of the process.

Leveled Flow Starts in the "Fuzzy" Front End: *Kentou* and Flow

Toyota's lean PD process attacks waste from the very beginning. A previous chapter emphasized how important front-loading is to leveling the workload with LPDS Principle 3: *Front-load the product development process to thoroughly explore alternatives.* Portfolio management, cycle planning, and rigorous shared resource scheduling at the front end of a process are prerequisites to leveling work in a multiproduct lean development system. Crucial to Toyota's ability to create flow during the execution phase is the ability to use cross-functional activities to front-load failure-mode resolution and core-engineering strategies, and align CE objectives during *kentou*. As previously noted, the *kentou* phase isolates, manages, and minimizes much of the variation in product development, which allows Toyota to focus on execution.

By aligning objectives across functions and developing designed-in countermeasures, Toyota enables synchronized, cross-functional process flow and eliminates one of the mortal enemies of flow in product development—*unscheduled and late engineering changes.* These engineering changes disrupt the process, drive excessive cost, and negatively impact quality. Although Toyota strives for no engineering changes after final design release, some engineering changes in a product as complex as an automobile are unavoidable. To mitigate the negative impact of these unavoidable changes, Toyota coordinates and controls engineering changes in its process logic.

The Role of Process Logic

At a high level, process logic defines the tasks and the sequence of tasks required to create a new product and the step-by-step process description that generates the schedules. Process logic determines who will do what and when, and which decisions PD teams must make at each milestone in the product development process at a macro level. It makes no attempt to provide all the details of how the work is done, but it does provide the framework that coordinates all the various participants. The functional organization that best understands the process, creates, maintains, and owns the detailed work instructions. Centralized control is limited to a critical, manageable few. In fact, fewer than 200 one-page process sheets define macrolevel product development process requirements. These are fashioned in such a way that they reveal the health of the program and serve as an early warning signal for any possible issues. All participants are aware of these requirements and align activities accordingly. Process logic by itself cannot create flow, but when it is flawed, it drives rework loops, waste, and prevents flow from taking place.

Simple but common examples of flawed process logic include concurrent engineering sequences that are poorly synchronized. For instance, when manufacturing engineering engages in detailed work on unstable design data, inevitable design changes upstream will undo all manufacturing engineering efforts. Another example is engineering and test sequences that do not factor in enough time for test analysis and countermeasure development. Finally, process logic is flawed when it tries to force teams to make decisions or commitments before sufficient, specific information is available (Ward et al., 1995). Engineers, frustrated by the broken process logic, often develop work-arounds that add and drive variation in the PD system, which further inhibits flow.

In the Toyota PD process, scheduling starts with using *process logic* and *milestone requirements* to balance the needs of each PD program. First, Toyota staggers programs to keep resource demands level. Next, Toyota decouples high-level schedules for vehicle subsystems with different content and time requirements. The company also separates and designs schedules for power train, chassis systems (underbody), and top hat (upper body) that will converge later in the process. Toyota ranks or scales each new product, as well as the manufacturing requirements, according to the amount of content for each of these subsystems. The amount of content determines specific milestone-timing requirements for each subsystem, and changes within each subsystem then determine the high-level schedule the program will follow.

Toyota's approach to macro process logic is the essence of elegant simplicity. It provides centralized control without the waste associated with monstrously large traditional PD central schedules (which are usually too complex to follow accurately) and places ownership and accountability where it belongs.

Workload Leveling, Cycle Planning, and Allocating Resources

In most companies, product development is a cyclic environment, and trying to level workload can be a maddening experience. However, it is a critical component of effective utilization of resources and speed to market. Leveling workload must actually begin prior to the execution phase; in a lean PD process, it begins before the surface transfer. In fact, it begins with product portfolio planning and resource scheduling, which occur prior to execution.

Product planning is the process of reviewing the performance of a company's current product portfolio to determine where potential gaps and opportunities may exist and then determine which products the organization will develop. A cycle plan is a time-based output of which products the organization will develop and when. Since the purpose of a cycle plan is to determine PD system resource requirements, it is important to make the plan relatively stable. However, given the economic and competitive forces in which most companies operate, maintaining stability can be a major challenge. Fundamental market forces change rapidly, and to survive, companies must have the ability to react quickly. Cyclical periods of intense resource demand or little or no resource demand whipsaws a company's PD system. Scale only further exacerbates the challenge. The larger and broader a company's product portfolio is, the more challenging this problem becomes. Toyota has handled this challenge in several ways.

Using Common Platforms

In 1992, Toyota reorganized its platform strategy into vehicle centers. Each vehicle center maintains its own planning division consisting of 170 to 200 people (Cusumano & Nobeoka, 1998). This planning division is responsible for the following tasks:

- advanced concept studies
- product portfolio planning
- cost planning (including component commonality)
- resource allocation

By creating a *planning group* within each vehicle center and developing the plan around common vehicle platforms, Toyota was able to decrease the number of products and therefore the scope and complexity of the PD plan. In addition, combining these products into vehicle centers and the platform strategy has improved predictive accuracy and communication, resulting in fewer changes to the cycle plan once it has been finalized. (The specific organizational characteristics of the vehicle centers is discussed in Chapter 8.)

Staggering Vehicle Launches

Toyota then schedules engineering redesigns in its product portfolio to level the workload over time. For example, Toyota would not schedule all the vehicles in a vehicle center for a complete redesign in the same year. A simple matrix would have rows representing vehicles within a vehicle center and years as columns. Toyota typically schedules its major redesigns of one vehicle with minor facelifts of other vehicles so as to have a similar workload in any given year.

Ideally, Toyota engineers would like to stagger annual vehicle launches so that an equal number of vehicles are launched in each quarter. However, the laws of the marketplace often preclude this. Sales and marketing determine the ideal time to launch a vehicle, and engineering has little influence on this. Thus, engineering must deal with the fact that late summer and early fall are going to be crunch times with major demands to support vehicle launches in various plants. Kunihiko Masaki, former president of the Toyota Technical Center, explains:

> Launchings of existing models are fixed, in general, in a product plan. But Sales channels have strong power to determine when they want the vehicles launched. But a few months delay or advance of launching is sometimes possible. Staggering of vehicle development projects is done in the technical center based upon the *Kousu Yamazumi* forecast of workload vs. workforce. General Managers submit their own division's staggering idea to the Technical Administration department. Workforces of Toyota Auto Body, Kanto Auto Body, ARACO, Central, Hino and Daihatsu can also be brought in if the workload gets too high.

The Execution Phase of Product Development

The execution phase of product development is quite different from the *kentou* or study phase. In the front-loading phase, the PD teams anticipate, study, and resolve problems, completing such tasks as fundamental design decisions, identifying failure modes, designing in countermeasures, and setting cross-functional objectives. Once *kentou* is complete, detailed engineering, prototyping, and production tooling can begin. By the time it reaches this point, Toyota has made a full commitment to the product and has begun to invest significant sums of money in tooling and in its suppliers.

Because of this investment, it is financially critical to have a high-velocity PD process with radically shortened lead times focused on precise execution and smooth product-to-market delivery. Toyota's goal, from this point forward, is to optimize capital investment, match quick cycle-supporting or embedded technology lead times, make decisions closer to the customer, and react quickly to changes in the competitive environment. Creating flow by synchronizing product development activities is one of the most powerful ways to increase speed.

Cross-Functional and Within-Function Synchronization

The PD value stream consists of all work and functional expertise required to bring a product from the planning stage to launch. Creating process flow within individual engineering activities is necessary, but insufficient for creating flow. To avoid interrupting the flow as a new product moves from one organization or resource to another, the cross-functional module-development teams must coordinate and synchronize individual functional organizational activities. *Cross-functional synchronization* is even more important for the successful execution of concurrent engineering, where simultaneous activities can result in a lot of rework if not fully integrated. Effective cross-functional synchronization in a lean PD system requires a thorough understanding of:

- the details of how the work actually gets done.
- each participant's specific roles and responsibilities.
- key inputs, outputs and interdependencies for each activity.
- sequences of activities in all functions.

When the participants in an upstream process understand these requirements, they know what to deliver so downstream participants can accomplish their tasks. Conversely, downstream participants can adjust their process to maximize the utility of the information available from the prior upstream process as it matures. A review of a number of ways that Toyota synchronizes activities, both within and across functions throughout the development process to maintain flow, follows.

The role of the simultaneous engineers (SE) is a powerful mechanism for both cross-functional and within-function synchronization. As discussed in Chapter 4, SEs are responsible for specific parts early in the program until launch; in this capacity, they serve as lead manufacturing engineers. This role of shepherding parts through the entire development process removes the momentum draining hand-off of knowledge and parts that occurs in a traditional PD process. Simultaneous engineers mitigate the extra setup time that engineers would otherwise have to expend when moving from task to task to support continuous flow. The SEs are responsible for making timely decisions, communicating regularly with functional resources, and working to keep things progressing, particularly on the critical handoff between product development and manufacturing process development. They are adept at decision making, transferring knowledge, and coordinating the activities of people inside and outside of their home organization as they work on their respective set of parts. Utilizing the SE on the MDTs also builds accountability in the lean PD system.

Examples of Cross-Functional Synchronization

One of the best ways to synchronize cross-functional activities is to align and integrate them. This is what Toyota does early in the design process where simultaneous engineers, representing various manufacturing disciplines, work with designers and product engineers to codevelop feasible designs. By working together in teams organized around specific vehicle subsystems (Module Development Teams), such as closures or underbody, they are able to execute true simultaneous engineering or develop processes and product. This differs greatly from the practice of checking feasibility of an existing design after the fact, which of course creates mountains of rework.

Another example of cross-functional synchronization occurs in the die design process. Toyota designs dies from the outside of the part in. This process, combined with standardized manufacturing processes, allows Toyota to maximize the utility of early data because part type and size are known much earlier in the process than details on flanges and holes, etc. When combined with standard die components, such as nitrogen springs and die wear pads, a great deal of die design work can be accomplished before finished part geometry is available.

The Toyota PD process also uses specific product development process events, such as design reviews, prototype builds, and part coordination builds that are organized to synchronize cross-functional activities in real time. For the design reviews and prototype build events, engineers and suppliers must bring the designs, prototype parts, and test results to be reviewed. This provides opportunities to coordinate activities in real time. Part-coordination build events, in which part-by-part evaluation takes place during a slow vehicle body build, are timed to support die try-out activities. In all three of these events, key team members gather to exchange information and adjust activities as needed. If required, the engineers make on the spot, real-time decisions, and engineering changes are coordinated across the group simultaneously to maintain flow.

Creating Flexible Capacity

By using all the lean PD methods described in this book, Toyota has managed to create a relatively predictable and repeatable product development process. In fact, through methods like standardization and adherence to detailed schedules, Toyota can anticipate the peaks and valleys over the life of a vehicle program. A predictable and repeatable PD process allows the company to plan resource allocation. This does not mean Toyota can control workload and level resources over the life of the program—natural ebbs and flows make this impossible. However, because Toyota can anticipate these ebbs and flows, it can plan for extra resources at very specific times by using its *flexible capacity system* to add extra engineering resources when they are needed.

Toyota creates flexible capacity in two fundamental but strategic ways: 1) satellite companies and 2) flexible staffing. *Satellite companies*, such as Toyota Auto Body or Toyota Auto Loom (now called Toyota Industries), provide a capacity relief valve. These are wholly-owned subsidiaries that are fully versed in Toyota operating methods and standards and can take up all or the most critical portions of programs as required by cycle plan demands. Because they are as competent as Toyota's internal resources, these companies can step in and execute with none of the ramp-up time or transaction costs associated with traditional supplier arrangements. The practice of using satellite companies is consistent with the single queue, multiple-server condition, which is an efficient processing strategy (Hopp and Spearman, 1996), as well as with the flexible capacity strategies recommended by Loch and Terwiesch (1999) in other industries.

To achieve the strategy of *flexible staffing*, Toyota pools and shares various skilled technical staff. Although each PD program team has a group of dedicated, highly experienced engineers, these are augmented with technicians and tracers (those who now detail designs in CAD systems) from pools that are shared by multiple programs. In addition, Toyota uses specially qualified suppliers as a final tool for staffing flexibility (see Figure 5.7). In this way, teams add additional capacity only when and where it is required, creating a kind of just-in-time approach to resource allocation.

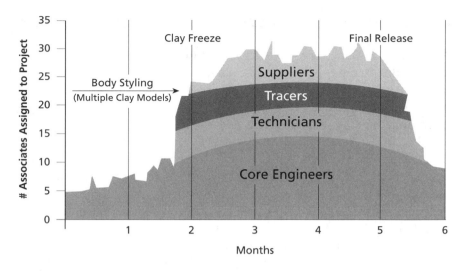

Figure 5-7. Toyota Flexible Staffing Distribution to Address an Unleveled Flow

While facilitating Toyota's phenomenal speed and product quality, standardization is also a crucial underpinning for flexible capacity in a lean PD system. Without rigorous standardization of skills, design, and process (or standardization cubed), you cannot achieve this level of flexibility because the learning curve of the flexible resources would be too steep. This *standardization-flexibility paradox* is discussed in Chapter 6.

Not to be overlooked is how standardization contributes to increased planning capability and subsequent system stability. When system variability is reduced, the time required to complete standardized tasks is more consistent and predictable over time. This reduces task-time variation. Tasks performed in a standardized process are completed on time, advancing the process according to schedule. This, in turn, reduces inter-arrival variation. As a result, PD programs are completed according to schedule, eliminating the need to expedite or otherwise change individual program schedules, which makes capacity planning far more accurate and stable.

Finally, Toyota's incredible speed to market is itself an advantage in stabilizing its cycle plan. The company's ability to execute to market demand quickly means it can target opportunities before they disappear and avoid the constant starting and canceling of programs inherent in slower organizations.

Detailed (*Fundoshi*) Scheduling to Head Off Unevenness

Accurate and disciplined scheduling is fundamental to managing multiproject workload leveling. As noted in Chapter 4, multiproject management is the ability to manage technological complexities and interrelationships while developing diverse and sophisticated products. Multiproject management optimizes the sharing of resources across multiple, concurrent projects. In a lean PD system, each individual PD program draws from the same set of designers and engineers; in addition, each utilizes the same tooling and prototype facilities. Obviously, if multiple programs vied for the same resources at the same time or did not proceed according to a planned cadence (missed intermediate schedule dates), a progress-blocking logjam would occur.

Designing and developing an automobile takes hundreds of engineers, thousands of components and tools, and high-tech equipment to manufacture each of those components. It also takes many specialized testing facilities to ensure quality, performance, and safety. Multiply this across all the PD programs running simultaneously, and you see the enormity of any automotive PD scheduling task. A completely centralized approach would quickly become too cumbersome and inaccurate, due to the many specialized tasks. A strictly functional approach would not provide the necessary cross-project functional synchronization and would drive PD teams to focus on optimizing their local objectives at the expense of the program as a whole. The challenge in scheduling within a complex environment is to schedule in only the details that accomplish the objectives—avoiding the waste of excess information and false sense of control.

At Toyota, schedule discipline means recognizing that intermediate dates are crucial to managing limited resources across multiple programs and approaching these dates with rigor and precision. Toyota's suppliers are also aware of this tenacious attentiveness to intermediate target dates and would not consider letting a date slip. Toyota executes to schedule at a level of precision that is unknown in traditional product development systems. All participants in Toyota's individual PD programs understand that these intermediate milestone dates are critical in order to manage the shared resources required by all programs. If they do not complete their requirements on time, they will not be given an extension and will have to "move to the back of the line." One example of just how strictly this is observed is that Toyota engineers are often prepared to sleep on tatami mats at test equipment facilities if it will help complete test cycles on time.

Detailed (*Fundoshi*) Scheduling at the Functional-Organization Level

Synchronized process logic and standardized milestone requirements allow the PD teams to concentrate on meticulous detail at the functional-organization level (e.g., prototyping, testing, or tool and die facilities). Toyota often refers to this detailed scheduling as *fundoshi* scheduling. *Fundoshi* is an ancient, traditional, male undergarment that is long and narrow and wraps around the body. When applied to scheduling, the term conveys something quite similar because *fundoshi* scheduling specifically refers to the *length* and the *comprehensiveness* of detailed schedules. At the functional level, teams have the necessary technical knowledge to create and execute these schedules in support of the high-level program requirements. The teams track all details of every part or tool. Specific machines and test equipment are scheduled and tracked by the hour. Tool and die components arrive JIT at the required work cells, and specialty cutters and machine programs arrive at the milling machines at the same time as the die components to be machined. Daily morning "walk arounds" at the facilities and large hour-by-hour schedule boards spell out specifc hourly task requirements, which provide the basis for visual "at the source" communication and identify issues early, addressing them on the spot.

Using Staggered Releases to Flow Across Functions

In manufacturing, it is easy to see the waste associated with working in batches—expensive inventory piles up, taking up needed space, wasting resources, and hiding potential quality problems. In product development, this is not as obvious—you do not see inventory piled up in the engineering cubes, but it is there just the same. For example, in a batch-design release system, manufacturing engineers wait for design information in order to begin their tasks. When a large batch of part designs is released that exceeds available manufacturing engineering resources, it sits in data collectors, waiting for someone to act upon, decide, or execute the next step.

In the lean PD process, program teams agree on a *design-release stagger*, which facilitates downstream operations. The release order facilitates work content so that parts requiring more time and tool-up in manufacturing engineering are released first and smaller components are released later. This practice also facilitates the design process. Because the larger and more complex parts are already designed, smaller components, such as brackets or reinforcements, can be designed in context.

Creating Process Flow in Nontraditional Manufacturing

Integral to the product development system is the nontraditional manufacturing required in prototype build and in tool manufacture. To make the entire PD value stream flow, these types of manufacturing tasks must move from unpredictable, craft-based activities to a nontraditional, lean manufacturing model. There are a number of examples of the application of lean principles within the die manufacturing activity. At the machining department of the die shop, for example, die components, cutter paths, and cutting tools are all required to complete the task of machining a die. Toyota's dies arrive at the machine just hours before they are required, as do machine tools that have been specially selected, sharpened, and kitted in rolling carts. The cutter path is pulled from a shared drive directly at the machine as required. Once die pieces are machined, they must be assembled, so the die pieces arrive JIT at standardized work cells, organized by construction task and die family. The same holds true for purchased components, which are assembled according to standardized and detail-specified work instructions. This creates a flow through the construction department that results in die construction times that are a small fraction of the time expended on similar tasks by Toyota's competitors.

During the tool-up portion of the lean PD execution, physical parts make inventory more visible. In a craft-based environment, however, inventory is often not recognized. Toyota staggers the release of die designs, foundry patterns, and castings as they are completed to create a flow through both their suppliers and their own tool shop. In the Toyota PD process, individual die designs are reviewed for function virtually and shipped to pattern suppliers as completed. Because of the accurate virtual quality, confirmation patterns are shipped as completed and individual castings arrive at the tool shop no more than two days before they are scheduled to begin machining. By staggering die by die releases instead of traditional batches, Toyota's tool process results in almost zero inventory and the fastest lead times in the industry.

Establishing an Engineering Cadence and Cutting Management Cycle Time

In lean manufacturing, takt time, based on available production time divided by customer demand, establishes the cadence and mix that drive manufacturing operations. Although the lean PD process does not have an exact corollary to takt time, establishing an *engineering cadence* to orchestrate activities and keep the program moving forward at a regular pace is crucial. Once you have eliminated waste and created leveled flow in the PD process, you need a mechanism to keep the entire system moving forward at a common, regulated pace. Toyota accomplishes *pace setting* by using a number of engineering cadence mechanisms. First, the company aligns and tiers engineering cadence with lower-level events designed to support higher-level program events. This keeps the program moving forward at a uniform pace and is achieved through rigorous design reviews, scheduled at regular intervals. Engineers and suppliers come to these reviews with prototypes, test results, open issues, etc., so that the CE can determine (at the source) whether the program is where it is supposed to be. The CE does not pull any punches and asks technical questions that reveal any potential concerns. Anyone who has attended one of these events knows that it is nearly impossible to fake your way through this process. You are either ready or you are not, and this becomes clear very quickly. Engineers work hard to prepare for these important design reviews, in itself a hard and fast deadline that creates subsequent deadlines for the team to meet. When issues or concerns arise, response is immediate.

Later in the PD process, the CE uses physical prototype build and part coordination events to the same effect. CEs schedule these events into the process to provide critical input into the design and tool manufacturing process at the moment it is needed to support the PD teams subprocesses. This again keeps the product development process moving forward. As with the design reviews, participants work hard to support these rigorous, system-focused events. The part coordination events, for example, focus on how stamped sheet metal panels fit together to create the car body and make individual die decisions based on this data.

At a lower level, the engineering leads meet with the CE in the Obeya about every other day to review program status, open issues, and performance to metrics, which are posted on the wall. These reviews are short and to the point; they focus primarily on any abnormal conditions, much like *jidoka* in manufacturing. In the Tool shop, the management team takes a daily morning walk around to check the status of parts to hourly schedules and develop countermeasures on the spot so that major events like prototype reviews are fully supported.

This integrated system of using cadence mechanisms has the added benefit of providing shorter management cycle times, which allow for quicker course corrections as needed and shorter time horizons for teams to focus on and work toward. Management cycle time can be defined as the time between checks on progress of the work by management and the amount of time to deadline for engineers. In a lean PD system the management cycle time is almost daily. In some companies, however, these management cycle times are so unstructured and vague that months may go by before someone sees the work results and notices problems. One automotive supplier's management cycle time was a classic example of this. After working with the authors, the supplier established a strict schedule for management cycle time and began holding weekly reviews, using visual boards to track progress. This decreased lead time significantly.

Using *Jidoka* and *Poka-Yoke* to Support Product Development Flow

In lean manufacturing *jidoka*, or autonomation, is the practice of recognizing an abnormal condition and responding quickly. Visual management, such as *andon (light goes on when a worker pulls a chord on the line)*, is often employed to aid in this effort, which is also associated with the separation of human work from machine work and is critical in supporting and maintaining flow. In the product development process, it is even more important to have the ability to recognize and remedy abnormal conditions quickly. The Toyota PD System uses a number of tools, such as checklists, for this purpose. Having program teams post specific objectives (visual management) in the Obeya and track program progress toward those objectives is an example of *jidoka* and visual management in the Toyota PD system.

Toyota's parametric CAD system is an example of utilizing technology to recognize abnormal conditions. In this system, as engineers update individual designs, all associated designs, including parts and most tool designs, are also changed to match. When an *error state* occurs in the system, such as part crashes, all appropriate designers are notified. In addition, Toyota creates all engineering designs within the context of the entire vehicle. This means all designers and engineers have access to all part designs simultaneously and can see if their planned change(s) will impact other parts.

Poka Yoke (error proofing) is another concept that supports continuous flow in Toyota's manufacturing process. *Poka Yoke* prevents errors from occurring, which reduces inspection time and frees up time to build in quality. A related concept is inspection at the source, which does not rely on outside inspection to catch errors downstream but makes the manufacturing operation responsible for checking quality as it produces. In product development, *Poka Yoke* takes the shape of:

- checklists
- standard, detailed test plans
- part quality matrices
- standard architecture
- shared components across vehicles
- standardized manufacturing processes

The checklists guide engineers through the PD process and eliminate design errors. Test plans, standardized part by part, prescribe test requirements and test timing for each part being designed. In addition, each part has a detailed quality matrix that identifies the standard manufacturing process and matches individual part design characteristics impacted by specific steps in the manufacturing process. The quality matrix (see Chapter 15) provides design engineers with process-driven design guidelines and builds in lean manufacturing. Adapting standard architecture to new products enables Toyota to maintain consistent vehicle performance levels relative to crash, handling, and noise. The same is true for sharing components across vehicle lines. Finally, by designing new parts to use standard manufacturing processes, Toyota can be certain of quality and productivity levels. A lean PD system prevents error states before they occur, which helps to create predictable results.

Pulling Knowledge Through the PD System

In lean manufacturing, pull production eliminates overproduction by having downstream activities signal their needs (demand) to upstream activities. *Kanban* cards usually signal (control) production in a pull system. In product development, knowledge and information are the materials that are required by the downstream activity The speed at which technology delivers information in automotive product development is overwhelming. However, not all information is equal to all people. The lean PD System uses "pull" to sort through this mass of data to get the right information to the right engineer at the right time. *Knowledge is the fundamental element (material) in product development.*

Toyota does very little "information broadcasting" to the masses. Instead, it is up to the individual engineer to know what he or she is responsible for, to pull what is needed, and to know where to get it. Individual engineers are expected to locate and extract needed information, whether this be design data residing in the data collector, a product performance experience, or a perspective from a senior executive. This policy holds true for everyone, from the most junior design and release engineer to the chief engineer. The key underlying principle that makes this work is that everyone has access to both the design data and the CE.

For an example from the opposite end of the program hierarchy, all engineers are responsible for creating benchmarks for their respective components. They are expected to gather relevant information and understand the latest technological developments, industry trends, and supplier and competitor products that affect their designs. Once the execution phase begins, manufacturing engineers pull design data from data collectors as they need itto start working on die or fixture designs. All engineers pull requirements from checklists, which are updated at the end of each program.

The supplier mentioned earlier (a company that had an unacceptable management cycle time) illustrates how the PD system links processes. This seat maker through value steam mapping identified that they were batch dumping information onto the next process (design sent hundreds of drawings to purchasing and ordered hundreds of parts prototyping, to build the hundreds of different variations of prototype seats, etc.) After moving to a staggered release system where a subset of seat designs were released on a preplanned schedule, weekly reviews of progress were set up and the supplier set up status boards at each functional area within the value chain. A key purpose of the status boards (referred to as "pull boards") was to signal the need for information from other functions. Once the status board was in place, it was easy to spot when key information was needed. When key information was delayed, it was identified within a week rather than months later. The example clearly shows that in a lean PD process, a key enabler for pull knowledge systems is reducing management cycle time.

Putting It All Together to Flow

Within manufacturing, the common solution to creating flow and thus linking together operations is the cell. Cells are created by taking equipment from different departments and arranging them in a product's natural flow. This will quickly move products one piece at a time through the cell. Where cells are not feasible, small inventory buffers are established with pull systems between processes. The general lean principle is to schedule in one specific place (the pacemaker process) and then pull material to that point. The pacemaker process is not scheduled exactly as orders come in from the customer, as customers demands are rarely level. Instead, production control levels the customer demand into a schedule in the pacemaker, which then creates a leveled pull on upstream processes and ultimately on suppliers. The speed of production at the pacemaker is set at the *takt* time—the average pace of customer demand over the period leveled.

The equivalent of a one-piece flow cell in the lean PD process would be a cross-functional team dedicated to a product design that works just as needed in the sequence needed. In a sense, this is what the simultaneous engineer and cross-functional MDT is doing. Outside information and materials can then be pulled as needed by this "virtual cell." Unfortunately, the level of precision of the routine processes in manufacturing is not possible within product development. However, as discussed throughout this chapter, Toyota has used analogous principles to drive out process variability and largely eliminate rework through rigorous task and schedule discipline, manage capacity and level workload through detailed planning and utilizing flexible resources in peak periods, create flow through staggered releases and cross-functional synchronization and orchestrate the entire system through cadence mechanisms.

As valuable as these mechanisms may be, a lean process strategy is insufficient. As should be obvious to the reader at this point, the execution of a lean process requires discipline and standardization. A concept that is covered in the next LPDS principle.

LPDS Basics for Principle Three
Create a leveled product development process flow

Utilizing a process perspective to improve product development performance is potentially very powerful. There are several critical characteristics of a lean product development process:

- Use the *kentou* or study period during concept development to anticipate and resolve as many downstream technical issues as possible, thus reducing variation early in the value stream.
- Develop a clear process logic with a manageable number of milestones and activities.
- Synchronize activities across functions.
- Level the workload through a well-designed product cycle plan which is adhered to in order to manage system capacity.
- Use a flexible capacity strategy to fill in the gaps in high workload periods.
- Use scheduling across functions, and even more detailed scheduling within functions, to synchronize activities and drive out variation.
- Stagger the release of data from one function to the next, prioritizing what needs to be worked on early versus late.
- Establish an engineering cadence and short management cycle time to orchestrate the system and create manageable deadlines.
- Execute the process plan with precision and demand schedule adherence to drive out inter-arrival variation.
- Use checklists and part-by-part standardized development plans to drive out task variation.
- Build in quality at each step of the process and do not pass along problems.
- Set up a system and culture in which engineers pull knowledge as they need it instead of inundating large numbers of engineers with information as it is produced.
- Build learning and continuous improvement into the basic process.

This is a long list, but the combination of all these methods will begin to develop a level flow and create a controllable process that companies can improve through *kaizen*.

CHAPTER 6

Utilizing Rigorous Standardization to Reduce Variation and Create Flexibility and Predictable Outcomes

"Today's standardization is the necessary foundation on which tomorrow's improvement will be based. If you think of 'standardization' as the best you know today, but which is to be improved tomorrow—you get somewhere. But if you think of standards as confining, then progress stops."

HENRY FORD

STANDARDIZED WORK IS ONE OF THE CORE DISCIPLINES of the Toyota Production System in which jobs are specified down to the second to match takt time—the rate of customer demand. The question here is whether this discipline can be applied to engineering work. Takt time standardization may lend itself to some routine tasks, such as the simplest CAD work, but engineers who move from big task to big task facing multiple uncertainties cannot standardize work in a way that specifies exactly what they will be doing every five minutes.

Indeed, when the authors have suggested to engineers that they need to standardize their jobs, the responses were predictable: "We are creative engineers," "We do not do repetitive manual work," "We need the freedom to schedule our work day and to be creative." To a certain degree, it is easy to understand why product development engineers are unable to perceive how standardization and creativity can work in tandem. On the other hand, Toyota's PD process shows that variations of standardization actually give program teams a great degree of flexibility and enables speed, precise execution, improved quality through robust reliability as well as system predictability, and waste elimination that reduces cost.

Standardization, coupled with a culture of discipline are the most powerful weapons a product development organization can bring to bear against the destructive power of variation identified in our previous discussion of queuing theory. In fact, standardization underpins and enables much of Toyota's success in product development. It is its very backbone. Rigorous design standardization supports the power of platform reusability, allows Toyota to share critical components, subsystems, and technologies across vehicle platforms, building in lower cost and higher quality. Standard architecture enables consistent body system performance, minimizes test requirements, and underlies consistent lean manufacturing processes. Standard development processes build trust, enable development speed through precise synchronization and are key to successfully managing the very complex process of developing new vehicles. Standard manufacturing and testing processes enable consistent quality and excellence in execution of lean manufacturing as well as make clear the upfront constraints on product development. Finally, standard engineering competencies ensure Toyota's ability to consistently develop outstanding engineers, produce consistently high levels of product development process performance, and are the basis for professional trust and collaboration. Far from diminishing the autonomy and creativity of engineers, when coupled with Toyota's pursuit of perfection, standardization is the very basis for a level of professionalism, pride, and an invigorating environment of technical collegiality and mutual respect unmatched in their industry.

Three Categories of Standardization

As noted in Chapter 4, there are three broad categories of standardization in the lean PD systems: design standardization, process standardization, and engineering skill-set standardization. Each category is briefly defined below and then analyzed in the context of Toyota's PD process.

1. *Design standardization.* This is standardization of product/component design and architecture. It includes the use of proven, standard components shared across vehicle models, building new model variations on common platforms, modularity, and design for (lean) manufacturing standards that creates robust, reusable design architecture.

2. *Process standardization.* This involves standardizing tasks, work instructions, and the sequences of tasks in the development process itself. This category of standardization also includes the downstream processes that test and manufacture the product.

3. *Engineering skill-set standardization.* This is standardization of
   skills and capabilities across engineering and technical teams. It is
   based on a deep commitment to people development and growth
   through demonstrated competencies. It is quite powerful and
   often overlooked.

Category One: Design Standardization and Engineering Checklists

Interestingly, many of Toyota's design standards are not given as specific
parameter requirements or "thou shalt or shall not" directives. More typically, these standards are concerned with ratios and physics driven. Sort of
"if, then" statements based on proven physical realities that give Toyota
engineers a great degree of latitude and creative freedom while simultaneously maintaining lean manufacturing requirements. Engineers are not
constrained by "point-based" parameters; instead, design standards provide a reliable guide as they work to identify optimal sets of solutions.

Design standards are embodied in specific, detailed part and process
checklists, reusable components and standard subsystem, and vehicle level
architecture that defines sets of best cross sections (often referred to as
common architecture) for each part. The power of common architecture
strategy was discussed in Chapter 4. The discussion below addresses the
use of engineering checklists as a key tool for design standardization.
Chapter 15 elaborates on this concept through specific examples and a discussion of trade-off curves that provide a graphic representation of how
they are used by engineers to achieve desirable solution sets.

Engineering checklists are certainly not unique to Toyota. In all prob-
ability, such checklists came to Toyota with the aerospace industry chief
engineers the company recruited as Japan's aerospace industry was
declining. (See Chapter 7.) Checklists are simple reminders of things that
should not be left out. They can be powerful or worthless, depending on
how they are used. If updated regularly and referred to diligently, they are
powerful. If stagnant and unused, they are worthless. Unfortunately,
many companies have not developed the discipline to maintain or utilize
them effectively.

Ideally, engineering checklists are an accumulated knowledge base
reflecting what a company has learned over time about good and bad design
practices, performance requirements, critical design interfaces, critical to
quality characteristics, manufacturing requirements as well as standards

that commonize design. Checklists at Toyota are highly visual, part specific, incredibly meticulous and comprehensive and may, in fact, seem impossibly detailed to those unfamiliar with the level of precision Toyota expects from its engineers. Checklists for complex parts may include hundreds or more parameters. Interestingly, as detailed and technical as the checklists are, most engineers we interviewed knew not only their own part checklists, but those of related parts and the checklists for the associated manufacturing processes by rote. This was clear evidence of consistent, long term use and a sense of ownership.

Though based on science, the real world practice of engineering is an art form that relies on tacit knowledge gained through experience and judgment in considering multiple variables that interact in complex ways. As a result, a best solution cannot necessarily be predicted in advance. It is learned over time through experience and is guided by the spirit of *kaizen*, which postulates that there is always an opportunity to learn more and that learning is an ongoing process. This spirit of engineering *kaizen* is driven by the never-ending pursuit of technical excellence that underlies consistent checklists utilization, validation, and improvement.

A company that cannot standardize will struggle to learn from experience and is not truly engaged in lean thinking. Indeed, any company that simply tries new things without standardizing along the way is "randomly wandering through a maze," repeating the same errors, relying on little more than undocumented hearsay and a wide range of opinions among its employees only to eventually discover that "it has been here before." Toyota uses a systematic and scientific approach to product development. It tests, evaluates, standardizes, improves, and retests, scrupulously following the Plan-Do-Check-Act cycle that was introduced to the company decades ago by Deming. It then standardizes "today's" best practice. As it accumulates new information and new experiences, these are used to modify current shared standards and reborn as a future "today's" best practice.

Toyota utilizes standards-embedded checklists from the very start of the program during the styling process through to launch at the assembly plant and everywhere in between. In the studio, designers and senior engineers from Body Engineering and Production Engineering work collaboratively through part-by-part solution sets, utilizing design and process checklists as their guide until a fully feasible design emerges from the process. In this way Toyota is able to design a feasible product the first time unlike NAC where engineers typically reviewed nearly completed designs on an ad hoc basis, virtually guaranteeing downstream engineering changes. In the launch readiness phase checklists are utilized to assure die and tool accuracy and to be certain that tools, dies, and manufacturing equipment are capable of maintaining critical part and assembly characteristics.

What this means is that Toyota has a plethora of checklists that, individually and collectively, reflect every part, every rule, every standard way of processing a given part, and every graph that illustrates acceptable and unacceptable ranges. In connection to PD, each Toyota engineer has books of standards that are checked off as each item for each design is considered. When explaining this structured and systematic checklist process during seminars or classes, the authors have heard a number of interesting and disturbing comments. In a classroom setting, hands instantly go up and invariably, someone will pose a variation of the following question:

In this day and age wouldn't it be better to computerize the checklists? We heard about the Toyota approach, and we are planning on doing them one better and developing an online knowledge management database of all the standards cross-referenced on a secure corporate intranet. We have a department set up to develop a state-of-the-art knowledge database. Is Toyota moving in that direction?

This is all rather moot. To begin with, Toyota has already moved in that direction and has computerized most of its checklists and standards. Secondly, creating a computerized knowledge database is not a guarantee of success, and in fact, misses the point: *you still might find the database is worthless.*

In essence, the question of how checklists are constructed and maintained, whether in books or in a computerized database, is secondary. The primary questions should consider the issue of roles and responsibilities. Who will feed the checklist? Who will use it? What are the specific responsibilities and accountabilities for updating the checklist and using it? At Toyota, this is intimately connected to an organizational structure (described in Chapter 8) that operates on the basic principle that "team work is the key to getting high quality work done but some individual always needs to be responsible." Responsibility for checklists is vested in functional groups that are organized down to the subsystem level.

For example, the door engineering supervisor is responsible for maintaining the door engineering checklist and ensuring that it is used by all door engineers. Body engineering and production engineering share responsibility for door checklists. A production engineering supervisor is responsible for checklists on how the door will be processed and the design features that make it producible. At the beginning of a new program, the door engineer will ask for the latest checklist from production engineering and considers it in tandem with the checklist from body engineering. Incorporating the production-engineering checklist into the design is now the door engineer's responsibility.

In short, the *people doing the work are responsible for maintaining and using the checklists and, in lean thinking, it is never a corporate IT function.* The checklist is not the amorphous responsibility of "engineers." It is the responsibility of a *specific* engineer responsible for each part of the vehicle, who must coordinate the efforts of all engineers working on that part of the vehicle and incorporate that knowledge, information, data, learning— or whatever you may want to call it—into the checklist.

As discussed in Chapter 5, Toyota's flexible capacity strategy relies on wholly-owned subsidiaries and skilled technicians who work out of pools arriving JIT to the product program. Rigorous process and design standards allow both subsidiary engineers and technicians to ramp up quickly and become almost instantly productive on the program. Because these technicians specialize by part, they are very familiar with relevant standards. They are able to apply the checklists, master cross sections, and standard locators to the design space, which is provided by the surface scans from the clay, and K4 body structures drawings, to produce final designs that reflect the new styling and performance intent of the new program. At the same time, they are able to retain proven part geometry that will maintain performance levels for such things as crash, NVH, and, of course, manufacturability. This reduces the number of physical prototypes that need to be tested. It also leads to far fewer late and expensive engineering changes, driving a lot of waste from the PD process. Having highly experienced engineers and technicians working with rigorous standardization tools is fundamental to Toyota's ability to deliver high quality designs rapidly and manage its flexible capacity strategy. A number of checklist examples are provided later in this book.

Category Two: Process Standardization

Using this second category of standardization enables true concurrent engineering and provides a structure for synchronizing cross-functional processes that enables unmatched vehicle development speed. A standardized development process means standardizing common tasks, sequence of tasks and task durations, and utilizing this as the basis for continuous product development process improvement. Process standardization is a potent antidote to both task and inter-arrival variation discussed in the previous chapter. Process standardization is the only way to know reliably what other functional organizations are doing and when they do it. It is how interdependent processes/organizations know specifically what inputs are required from each other and when they are needed. Finally, strict process discipline coupled with standard development processes are the only conceivable way to run a multiproject "development factory" and is absolutely fundamental in gauging the performance and progress of any individual program

As mentioned in the discussion in Chapter 5 on process logic, the lean PD process centrally controls high-level standard process requirements to guarantee synchronization. For Toyota (see Chapter 4), this means that macrolevel milestones and timing are utilized across different programs and that each individual functional-organization level controls the detailed, working-level processes. By leveraging both of these standardized structures, detailed, program-specific schedules at the working level are developed.

By contrast, NAC uses a corporate staff group to standardize milestones at a relatively high level with a great deal of detail about all functional organizations results that must occur by these milestones (e.g., "stage-gate model"). It is the responsibility of the program and functional teams to figure out how to accomplish this. Whereas Toyota's various engineering organizations each standardizes the means to engineer the product based on the requirements of a central framework, NAC corporate staff attempts to standardize only the ends for the entire product development enterprise.

Without rigorous standardization and common architecture, NAC lacks an effective flexible capacity strategy, resulting in constant bottlenecks at critical resources throughout the PD process. NAC does use supplier engineers or outsource engineering work, but because it does not standardize skills, design, and process, it often experiences poor results and extremely high transactions costs which it blames on their suppliers.

Having standardized processes guiding detailed work at the functional level is key in enabling flexible capacity and leveling workload. Without it, LPDS Principle 3 of creating leveled flow would not be possible in product development. It is well known in lean manufacturing that stability is a requirement for flow and this is also true in product development. Standardization provides the stability, consistent expectations and level of predictable outcomes as the necessary foundation for flow.

One perhaps overlooked benefit of a standard development process is that it contributes to more precise communication and greater understanding across engineering organizations by providing a common framework for discussion.

Toyota's Standardized Process for Production Engineering

Because of a reliable standard process, at the same time Toyota body engineers begin to work with digitized surface data from the clay model, production engineering is able to start its work on detailed process design. Die design, fixture design, and processing/binder development groups also begin preliminary activity at this time. The various Production Engineering organizations synchronize their activities with the evolution of the part design to maximize the utility of sketchy and partial design information to create a synchronous evolution. By utilizing a standardized process that works only with stable aspects of the part design as they become available, the Production Engineering Department creates efficient process flow through concurrent engineering and simultaneously eliminates wasteful downstream rework. This highly synchronized approach to concurrent engineering is vital to protect against the all too common practice of trying to accomplish too much too early with partial or premature design information, which is likely to change and cause rework and waste. In addition to the ubiquitous checklists discussed earlier, Production engineers also employ *senzu*. *Senzu* are very detailed manufacturing drawings that have been created for each part. *Senzu* are updated at the end of each program, shared across functional specialties, and contain all manufacturing information, best practices including manufacturing geometry changes, locators, weld locations etc., accumulated for a specific part.

Toyota's Die Engineering

During this intense period, the die engineering group practices a flexible capacity strategy, which includes the use of trainees. Because the lean PD process has broken down the complex die engineering challenge into many standardized subroutines, and because solids databases provide standardized components and simultaneous access to the designs, using computer-aided die design, designers with varying degrees of expertise can work on the same design at once. As a result, designs are completed more quickly and flexible human resources can then be transferred to work on other parts as they become available. Parts availability is especially important because in order for body engineering to achieve final data release requirements, some parts will be completed before others. Furthermore, because these parts are available to production engineering through a shared database, they can be *pulled* as the engineers are ready to work on them.

As noted above, cross-functionally synchronized standard processes enable die designers to start on parts with incomplete data. As each part design is completed, die designers can *pull* and complete it. At this point, die design trainees can carry out the die designing tasks that require less skill. These trainees then move on to assist another die designer. This process is possible only because standardization mutually supports all three lean PD subsystems—process, tools and technology, and people. One of the key enablers in establishing this kind of standardization is the *senzu*. In the case of dies, the *senzu* details such things as binder shape, overcrown, overdraw, and radii requirements. Collectively, *senzu* are part and vehicle specific references that are critical to stamping engineering performance in the entire PD process.

Process and Binder Development

Early in the process, in conjunction with the final phases of vehicle styling, standard manufacturing processes and common part architecture enable preliminary binder development to be accomplished in the processing/binder development area. The binder is the part of the first forming die that holds the sheet material in place while forming the part. It is crucial to a quality stamping process and can be quite challenging for complex geometries. This process often involves formability assessment, utilizing formability simulation and Finite Element Analysis (FEA). In fact, because there is not process or design standardization, NAC is forced to perform FEA on all stamped parts, creating a huge bottleneck in the process at this limited resource. However, in the Toyota PD process, characterized by standardized part geometry and manufacturing processes, less than one-third of the stampings require FEA of any type. Eliminating the need for FEA for two-thirds of all parts removes the potential for bottlenecks and associated queues and variability from the PD process. It also improves average throughput times and significantly reduces costs.

This also strengthens competitive advantage in the lean PD process of the next product development phase.

Toyota's Lean Tool and Die Manufacturing

In this phase, in which tools are cast, machined, assembled, tried out, and approved, Toyota's lean manufacturing principles are truly brought to bear on the product development process because building tools and dies is a form of manufacturing. Extremely accurate and detailed standard die designs allow Toyota to employ an adapted form of lean manufacturing principles at this point in the PD process that is a powerful competitive advantage in a lean PD system. On average, a large set of dies, such as those required for body sides, will require less than four months for casting, machining, construction, and preliminary tryout at the die shop. Because they are still utilizing a craft-based process for die making, most of Toyota's competitors require 10 to 12 months for the same set of tasks. The die shop will then ship them to the stamping plant for home-line tryout, which includes not only final die tryout but also tryout time for all the check fixtures and automation required for final production stampings. This takes up to an additional one to two months, in increments of six to eight hours once or twice per week. Toyota's high velocity die development capability combined with part design standards and effective use of sophisticated virtual tools has also allowed them to eliminate the need for most prototype tooling—a huge cost and time saver. Although Toyota has significantly faster times on special projects, the discussion below focuses on the typical time frame it takes for designing dies and completing tooling in a standard process.

Typical Time Frames for Lean Tool and Die Manufacturing

Designers classify all dies in categories ranging from *A0* to *D*. Assigned to each category is a specific line of milling machines, construction bays, and spotting presses. Those who have studied TPS will recognize this as the identification of *product families* with dedicated flow lines. *A0* are large dies with class one (outer) surfaces, *A* dies are large dies with unexposed surfaces, and so on to *D* class dies, which are smaller parts produced on progressive dies. Dies are assigned to the smallest possible line of equipment (right sizing). These die categories allow for standard procedures and standardized times that make scheduling more accurate and outcomes more predictable. To assist in visual management and to make all participants aware of scheduling requirements, the plant keeps and maintains large schedule boards that are both task and department-specific. These boards track progress in hourly increments and are checked by the plant management team daily. Once again, the lean standardization in the PD process works to reduce process variability and thus improve throughput times.

Designers perform virtual checks on the die designs against standards and then uplink the design to a virtual simulation system (digital 3-D simulation of the complete die set), to ensure clearances, functionality, and production efficiencies. These designs are then utilized to drive standardized work instructions in the die construction phase and are the models used by manufacturing to machine die patterns. This step eliminates the need for physical reviews. Toyota pattern timing is one week or less and die castings require only an additional ten days. In contrast, Toyota's North American competitors require three weeks for patterns and four weeks or more for castings.

Toyota Die Machining

Toyota's accurate and highly detailed die designs and standardized die manufacturing process allows the company to do the vast majority of die manufacture on precision milling machines, which substantially reduces time spent on handfitting and reworking die details, and in lengthy die tryout, advantages Toyota's North American competitors do not have. Toyota has also patented a number of specialized cutters to maximize the efficiency of its machining operations, adding an even greater element of precision, speed, and predictability to its lean die manufacturing. By focusing on precision machining, Toyota has completely eliminated several secondary operations, such as the hand polishing and die fitting required in traditional die manufacturing. This lean die manufacturing approach allows Toyota to apply additional lean methodologies such as SMED (single minute exchange of dies) to its machine setup operations and JIT cutter kit arrival to maximize machine value-added time. Detailed schedules in hourly increments are posted next to the machines and maintained by the operators. As in pattern construction, during morning plant walk arounds, the plant manager and the team review these schedules.

Toyota Die Construction

After machining, each die detail or component is shipped to the appropriate construction cell in the construction bay corresponding to its category classification. There are five separate construction bays, each containing several phased "lines" made up of multiple stations or cells, each of which is responsible for a portion of the construction process. Die details and purchased components arrive at the right cell at the right time to keep the construction process flowing forward. Just as in an assembly line, each cell completes a portion of the construction work, with the die moving on to the next cell for the next step in the process. Work procedures are thoroughly standardized in detail and each cell's task times is equalized (as measured in days) so that there is synchronized movement or flow through the construction department. This reduces variation, makes outcomes much more predictable, and enables people to see die construction status at a glance. Schedule boards are posted throughout the construction bays and all participants are aware of scheduling requirements and work to meet them.

Each cell is self-contained with all the assembly tools and perishable supplies (such as screws and dowels) that are needed in a particular operation organized and located near the point of use. Die locations are painted out on the floor, all hand tools and machines are in place, and benches with labeled racks and drawers are set around the working area. Air tools are suspended by retractable lift assist equipment; they are within easy reach when needed and recoil out of the way when not in use. Even purchased components arrive just before they are needed, so die makers do not have to leave their cells to search for anything. Die construction personnel resemble race car pit crews each busily executing predetermined tasks simultaneously with incredible precision. *Kaizen* is ongoing. For example, one *kaizen* focused on a construction innovation that eliminated flipping over the die during assembly—normally a time-consuming operation that requires a crane.

Die makers are cross-trained in construction tasks and also in tryout. Individual die maker skill levels are posted on boards in the department. Pay is linked to skill level, and all die makers are on salary but paid for overtime. Die makers use customized checklists for each cell in each bay, both as a procedural reference and a quality assurance tool. These checklists serve as the primary guide for the cell leader's sign-off of each die before it moves to the next station. These "sign-offs" by construction personnel are the primary form of quality control during die construction. As with pattern construction, no paper drawings of die designs are required: All die makers are trained to use the CAD system, and all die design data is available on the CAD computer located near the cell. Die designers and simultaneous engineers visit construction bays regularly to work with die construction personnel to identify problems or ways to improve designs and methods.

Toyota Vehicle Assembly Engineering

Vehicle assembly engineering is a part of the production engineering organization and is fully integrated with the die engineering group at Toyota. Their challenge is to design tools and processes to assemble the stampings into a vehicle body or body in white. This is a complex process, requiring large stampings to be precisely located, held in place, and usually spot welded together. Closure subassemblies such as doors, hood, and lift gates have an additional step of hemming in which a flange from the outer panel is hemmed or crimped around the flange of the inner panel in order to secure them together before being mounted on the vehicle. This requires the design and manufacture of complex and extremely expensive fixtures, subassembly cells and body assembly lines.

By standardizing locators, checking points, weld standards etc., Toyota was the first to be able to design stampings and assemblies to standards that support flexible assembly operations that allowed Toyota to build multiple body styles on the same line. In a more recent development to that process Toyota introduced Global Body Lines through their Blue Sky project that took flexible body assembly to a new level. This project required intense collaboration between both Production Engineering and Body Engineering to update design and process standards to support this revolutionary innovation. According to Atsushi Niimi, former President and CEO of Toyota Manufacturing North America, the new system replaces the fifty pallets required for each body style in the old system with only one master pallet tool each. This new pallet tool looks something like a ski-lift and locates the body from the inside on programmable locators. This system improves over-all body quality, reduces the number of weld stations required, and dramatically increases manufacturing flexibility. Now eight different bodies can be assembled on the same line by changing only a single master pallet. Niimi-san claims that new body shop installation costs for new programs have been reduced by 50 percent, space required is reduced by 50 percent, and the cost of adding another body to an existing line or a new top hat program is down by 70 percent. This is the power of innovation. But it would not be possible without engineering organizations working together collaboratively to create effective process and design standards and a culture of discipline to maintain the gains.

Toyota has installed the GBL worldwide. Consequently all vehicles are now designed to support this standard assembly process.

Category Three: Standardized Skill Sets/Competence

Most companies considering standardization seldom think of standardized skill sets. Yet this is an essential principle for creating a lean PD system. It builds team integrity, enables incredible development speed, and drives task variation out of the development process. Managers have much greater flexibility in assignments and both managers and team members alike can have more confidence in performance expectations. Toyota's culture of demonstrated technical excellence is fundamental to creating professional trust and high-performing teams in any environment. This section of the chapter highlights some of the practices that lead to consistent or standardized skill sets at Toyota, beginning with the hiring process. A more detailed discussion of the benefits of the people development process is presented in Chapter 9, which deals with the seventh LPDS principle: developing towering technical competence.

After a lengthy and rigorous review process, Toyota hires only about 1.1 percent of professional candidates applying for engineering positions (Kramp, 2001). Once hired, engineers follow a standard, skills-acquisition-based personnel development process from day one. The process focuses on demonstrated competencies and intensive technical mentoring for advancement: A rookie engineer can expect to undergo an intensive two-year on-the-job training period before moving up to first-level engineering rank. Toyota invests three to four years in each new engineer before he or she becomes a serious team contributor. Within the industry, this is a significant investment. After this initial period, a body engineer can expect to spend five or six more years within this same technical specialty before being considered a first-rate engineer. During the approximately eight-year development period, engineers are "interviewed" four times per year, and technical areas of improvement are assessed using *standardized skills inventories*. Training is mostly on the job and special care is given to the assignments that an engineer receives to be certain he or she will have the opportunity for continued technical growth. An action plan is developed through *Hansei* (reflection) to address shortcomings. Among the criteria used to evaluate Toyota engineers is successful adherence to process and standard methodology, which further develops each engineer's standardized skill sets.

A new engineer's career path consists of experiences that develop deep technical competence, while slowly climbing the technical hierarchy within each functional department, and is a direct result of engineers being rewarded for technical achievement. The engineer's boss usually knows how to do the job better than the engineer; he or she also knows the standardized process for doing it, which enables the leadership principle of teaching and mentoring. The lean PD system depends on mentoring for developing talent. To support mentoring, Toyota creates an engineering apprenticeship environment in which highly technical, tacit skills are handed down from one generation to the next, thus basing professional growth on demonstrated competence in the real world.

Conclusion

This chapter concludes the discussion of the first LPDS subsystem *process* and its four principles within the broad framework of the product development system, which was outlined in Chapter 2 as a sociotechnical system (STS) with three primary subsystems: 1) process, 2) people, and 3) tools and technology. In STS terms, body engineering was used to show the technical system of processes—all the tasks and sequences needed to bring a body design from concept to start of production. In developing this section of the work, we emphasized how raw material consists of information, customer demands, past product characteristics, competitive product data, and engineering principles that are transformed through the lean PD process into the complete engineering of a product. Similar, if not stronger, emphasis was placed on the connection between lean development and lean manufacturing. In addition, the authors have endeavored to show how the first LPDS subsystem and its principles define a company's value stream map as how information flows, stops, gets rerouted, and sits in queues. The next chapter examines the second lean PD subsystem, *People*, and LPDS principles five through ten.

LPDS Basics for Principle Four

Utilize rigorous standardization to reduce variation and create flexibility and predictable outcomes

In a lean PD system, you need to standardize products, processes, and competence to create a foundation for flexibility and speed. Standardization is critical to the LPDS because it underpins many of the other LPDS principles by reducing variation, subsequently creating greater flexibility and more predictable outcomes. In LPDS, there are three types of standardization: design standardization, process standardization, and skill-set standardization, all of which are necessary to drive out waste and achieve a truly lean system. Design standardization is manifested in engineering checklists, standard architecture, and shared/common components and platforms. Process standardization refers to both the development and manufacturing processes and is housed in individual component development plans (*senzu*) and detailed manufacturing process plans. Standardized skill sets are developed through careful mentoring, strategic assignments, and periodic assessments of demonstrated competencies. The functional organizations own, maintain, continuously improve, and execute standardized designs, processes, and skill sets.

SECTION THREE

People Subsystem

CHAPTER 7

Create a Chief Engineer System to Lead Development from Start to Finish

We can be successful at Toyota only when we do something better than our competitors or when we surpass the average for the industry. If we are designing a new product, and know there is no room for failure, our attitude certainly must not be just to aim for the average. If we do that, we will surely fail. We must do our all-out best at such times, and allow ourselves no thought of failure.

KENYA NAKAMURA, first chief engineer of Toyota Crown

THIS CHAPTER BEGINS A DISCUSSION of the second STS subsystem for the Lean Product Development System, *People*, which cover Principles 5 through 10. People are the essence and energy of the LPD system, and you cannot compete in product development without a capable, energized, aligned organization that executes as a high performance team. Lean PD people systems are constructed on a foundation of teamwork, continuous learning, and *kaizen*, which drive and evolve lean within the other subsystems. The *people* subsystem encompasses shared language, symbols, beliefs, and values that determine such things as how an organization is structured, its leadership style and learning style, and how it recruits, trains, and develops employees.

Consider what it would take to align your company with the LPDS model's people system. Would you need to overhaul your business and managerial philosophy? Can you really change the "soft" thing known as "culture?" The answer is complex and entails changing some fundamental approaches to how people get work done. For the lean PD system, the first important change is deciding who "runs the program"—you have already been introduced to the person who does this at Toyota: the chief engineer (CE). Determining this piece of your lean PD system is one of your most important decisions because it dictates how you develop the structure of your PD organizational system.

Most organizations use some form of the matrix system to determine who is reporting to whom and what respective roles and responsibilities will be. (Toyota's Matrix System is discussed in detail in Chapter 8.) Within this matrix, each engineer has multiple bosses—typically a functional boss who is a technical specialist (e.g., Body Engineering) and a program boss who runs the product development program. In most organizations, the matrix is a communication nightmare that spawns mixed allegiance and conflict between and among different parts of the organization. Toyota also uses a matrix, but the program boss is the chief engineer. The CE as program boss system is unique and avoids many of the problems within a traditional matrix structure, mainly because the CE clearly runs the show. Thus, it is appropriate to begin the discussion of the people subsystem with the central figure of Toyota's product development organization.

The Cultural Icon Behind the CE System

Toyota's chief engineer, sometimes characterized as the "Heavyweight Project Manager," is likely the most publicized and imitated aspect of Toyota's PD system. But while it is true that most organizations have some type of program manager who, like the CE, is responsible for overseeing design projects and making sure they are on time and on budget, the similarities end there. Like the program manager in many other companies, the CE does not have formal authority over the engineers working on the program; however, the CE is ultimately the one person charged with the success of the design, development, and sale of a car. Consider the range of responsibilities for the CE and his or her small staff:

- voice of the customer
- customer-defined value
- product concept
- program objectives
- vehicle-level architecture
- vehicle-level performance
- vehicle-level characteristics
- vehicle-level objectives
- vision for all *functional program teams*
- value targets
- product planning

- performance targets
- project timing

The CE's ultimate responsibility is delivering value to the customer. While Toyota always emphasizes teamwork, there is always one person who is accountable for the success of the team. For product development, this person is the CE.

The defense industry in Japan originally used the CE system, and like many other process innovations, this system was adopted and adapted by Toyota. The geniuses given credit for building the CE system at Toyota include early chief engineers like Tatsuo Hasegawa, chief engineer for the first *Corolla*, and Kenya Nakamura, chief engineer for the first 1955 *Crown* (Ikari, 1985). Since that time, there has been a continuous stream of CEs. But while the tradition has grown, the responsibilities and characteristics of the chief engineers remained fundamentally unchanged. Below are some of the characteristics that Toyota has come to value in its CEs, many of which underscore the importance of balancing engineering skills with a broad and rounded approach to leadership:

- A visceral feel for what customers want
- Exceptional engineering skills
- Intuitive yet grounded in facts
- Innovative yet skeptical of unproven technology
- Visionary yet practical
- A hard-driving teacher, motivator, and disciplinarian, yet a patient listener
- A no-compromise attitude to achieving breakthrough targets
- An exceptional communicator
- Always ready to get his or her hands dirty

While working with companies on lean product development and introducing the CE system, the authors often hear, "We already have a chief engineer" or "We can tweak the organizational chart in this way to create this role." The response to such comments is that although it is easy to imitate the organizational form of Toyota's CE system, it takes years to develop the roles and responsibilities within that system. Furthermore, the lean PD model requires careful selection and grooming of CE candidates over many years (at Toyota, this grooming period takes at least 12 years and usually more). This grooming entails an incremental blending of characteristics that are the hallmarks of a super engineer *and* leader.

First, management needs to identify exceptional people who have handled challenge after challenge, have matured and grown, and have the hands-on experience required to lead the development of a highly complex product. Second, management needs a CE to fit into a broader system of roles, responsibilities, and commitment. The CE role is high profile; it is a role the entire organization must recognize and respond to. In a sense, this person functions outside the company bureaucracy and standard operating procedures and has the latitude to do what is necessary to get the job done. The position is valued and is often more admired than the organizational senior positions of director or vice president. It is by far the coveted position among engineers within the Toyota PD community. For all of these reasons, the chief engineer at Toyota is revered as a cultural icon, one whose special status lies in the freedom to rise to many challenges.

In a typical PD system, once the top executives make a strategic decision to develop a new product line, the program is immediately assigned to a high-level planning group, possibly with a marketing background, which develops the marketing concept. Industrial designers are then assigned to develop sketches. Somewhere down the road, engineers enter the picture to work out the technical details. A lean PD system does not take such a soft, non-technical route. When top management makes such a momentous decision around a new program, it immediately selects a chief engineer to head it up. As project manager, the CE does not simply coordinate schedules or fine tune technical details—the CE owns the car, from concept to styling, to prototype, to launch. Often, a CE will stay with the same product over multiple generations.

A Tale of Two Chief Engineers: *Lexus* and *Prius*

To appreciate the unique role of the chief engineer in the Toyota system it helps to consider two of the most visible cases of product development at Toyota over the last 20 years; two vehicle programs that Toyota views as the epitome of the Toyota Way.[1] For *Lexus*, the breakthrough approach meant developing not only a car but also a brand name with its own dealer network. *Prius*, on the other hand, was the first fuel-electric hybrid vehicle ever to be mass produced. Many people know that Toyota excels at rapid

---

1. In this section, we summarize the *Lexus* and *Prius* story line from *The Toyota Way* (2004) for the purposes of highlighting the PD process. For a more in-depth story and discussion around the personalities of the two CEs, the reader can refer to *The Toyota Way*.

vehicle modification or top hat programs rather than innovative sweeping change to meet changes in the market. However, fewer people know that Toyota began with a strong spirit of innovation when engineers had no choice but to be innovative. This goes back to the days of former Toyota president Shoichiro Toyoda and his "3 Cs"—creativity, challenge, and courage—and it was this spirit that permeated the development process of both the *Lexus* and the *Prius*, with the chief engineers leading the way. By featuring these exceptional programs we hope to demonstrate more clearly the critical role of the CE.

In the lean PD process, the CE is tasked in the earliest stages to lead in developing the concept of a new vehicle, while the top executive's role is to tap the right person for the program. Toyota's *Prius* and *Lexus* were special cases. Neither vehicle had any preliminary styling nor was there a well-thought-out vision. With the *Prius*, top executives had a concept of developing a high fuel efficiency car for the twenty-first century. With the *Lexus*, top executives had a concept of building a luxury car for the U.S. market that would be in the class of such luxury carmakers as BMW and Mercedes. In both cases, it was the responsibility of the CE to shape an amorphous vision and make it a clear and definite *product concept.*

The genesis for the *Lexus* started with Yukiyasu Togo, head of Toyota Motor Sales in Southern California, who strongly promoted the idea for developing a luxury brand, eventually convincing management that Toyota needed a new car, a new brand, and a new dealer network. Once top management approved the *Lexus* concept, they quickly appointed Ichiro Suzuki as the chief engineer, one of the best and most revered chief engineers in Toyota's history. For the *Prius*, Toyota took a very different tack, appointing the neophyte Takeshi Uchiyamada to run with management's general vision of a twenty-first century car and rethink the development process itself. Uchiyamada, a former test engineer, who had not previously served the company as a CE, was also tasked with defining and engineering the product concept, as well as with launching the vehicle under an aggressive timetable. The two engineers came from very different backgrounds and had very different styles, but top management chose the right leader for each project, for reasons that went far beyond developing a new car.

Lexus: A Chief Engineer Who Refused to Compromise

In August 1983, the board of directors, in a top-secret meeting presided over by Toyota Chairman Eiji Toyoda, officially blessed the *Lexus* program.

when appointed as CE, Ichiro Suzuki immediately sought out the *voice of the customer* by conducting focus group interviews in various locations, creating a sampling of two groups across the main European luxury brands. Group A had four Audi 5000 owners, one BMW 528e owner, two Benz 190E owners, and three Volvo 740/760 owners. Group B almost directly paralleled this makeup. Suzuki collected the comments from the customers and classified them. Starting with the broad current concept of a luxury car, Suzuki looked into the customer's reasons for purchasing or rejecting other vehicles in the same class, according to the image the customer had of each car. He then defined the most common and simple qualitative descriptors and created a one-page summary (see Figure 7-1).

Voice of the Customer—Customer Value Characteristics

	Reason for Purchase	Reason Rejected in favor of Competitor
Benz	Quality, investment value, sturdy	Too small; weaker style appeal (vs. BMW)
BMW	Style, handling, functional	Too many on road
Audi	Style, space, affordability	Poor quality, poor service
Volvo	Safety, reliability, quality, sturdy	Boxy styling
Jaguar	Most attractive styling	Poor quality, small interior

Source: Reprinted with changes, from Jeffrey K. Liker, *The Toyota Way* (New York: McGraw Hill, 2004), 44, by permission of McGraw Hill.

Figure 7-1. Reasons for Purchase and Rejection of Competitive Luxury Vehicles (1980s)

What may surprise many engineers is that the Toyota sales and marketing organization was not driving this process. Where were the pages of marketing surveys? Where were all the scientifically valid samples? How did a single CE think that reviewing information from a couple of focus groups was enough? Nonetheless, this is exactly what happened, and this early work of studying and classifying the voice of the customer into descriptors like quality, investment value, and affordability, helped define *customer-value characteristics* that went into building the luxury concept for the biggest Toyota investment of the 1980s. Though this luxury concept was highly qualitative and intuitive, Suzuki had correctly read the current environment in which *Lexus* was going to be competing. In a traditional PD system, management would have had another group do this research and then this group would have tossed it to engineering to design-in these qualities. The shortcoming with this traditional approach is that as a program proceeds, this early work easily gets lost or ignored because there is no authority behind it to continue championing and refining it. In the case of the *Lexus*, the main thing Suzuki had learned was that a Japanese car was not associated with luxury or status. Armed with this information, Suzuki now knew what had to be done, and it was clear that this would not be business as usual. If it was to be successful, the *Lexus* would have to break out of Toyota's existing product development mold.

Suzuki's next challenge was to define the *Lexus* concept by what would appeal to customers in the future, not simply reflect what current customers or marketing told him about current trends. For example, he had asked customers to rank *customer value characteristics* in their decision to buy a Mercedes in the current market, but then outright rejected the data. Below is the customer's ranking Suzuki compiled, ranked from most important to least important:

1. Status and prestige of image
2. High quality
3. Resale value
4. Performance (e.g., handling, ride, power)
5. Safety

Because he was an engineer, this ranking was at odds with Suzuki's common sense. Being *intuitive yet grounded in facts*, he did not believe that these customer value characteristics defined the future luxury car. Suzuki stepped back to define some basic principles for determining the *vehicle-level architecture* that would lift the *Lexus* above other luxury cars and dominate the market. He asked the follow questions:

- What does it mean to own a high-quality luxury vehicle?
- What characteristics does a car need to make people who own it feel like they're wealthy—financially and emotionally?
- What characteristics would a car have that, as the years go on, would make people become more attached to it and possessive of it?

After extensive discussions with the CE team and other associates, Suzuki concluded that the two most important design criteria driving the *vehicle-level architecture* and meeting *the customer-defined value* for the *Lexus* and the future luxury car was, in order of importance:

1. Exceptional functional performance (power, feel, noise level, aerodynamics)
2. Elegant appearance (not traditionally a Toyota strength)

Suzuki decided that if Toyota could make a car with exceptional functional performance, which was considerably better than what Benz cars offered and, at the same time, combine this with an elegant appearance, the company could break free of its stodgy image and compete in the luxury market. With the vehicle vision now clearly defined and the *vehicle-level performance goals* set, Suzuki had the beginnings of his CE concept paper.

This upfront definition of the concept of the vehicles may have been Suzuki's biggest contribution to the company and the *Lexus*. Once the concept was defined, he proceeded to develop a set of "no-compromise goals." These were competing design criteria that, at first glance, appear mutually exclusive. If engineers focused on exceptional functional performance, then most likely, they would sacrifice elegant appearance. Likewise, if the engineers focused on appearance, they might sacrifice performance. If the engineers attempted to design-in these two criteria at the same time, they would probably be making trade-off choices. This is where the Toyota engineer in Suzuki kicked in. His overarching goal for the *Lexus* was to fuse these two criteria together, so the vehicle characteristics needed for one would benefit the other. The challenge he put to the engineers was to solve each problem resulting from the clash of these competing engineering criteria and synthesize it into the *Lexus* brand.

To accomplish this Suzuki had to find the right people in the company to do the right thing. This was to be extraordinarily innovative engineering and would have to accomplish some extraordinary feats, for example:

- Breakthroughs in vehicle styling that also achieved aggressive vehicle aerodynamics goals.
- Breakthroughs in engine design and machining of engine parts to achieve the world's quietest engine, with the least vibration facilitating low noise yet high performance.

Making these breakthroughs was not simply a matter of making a request and sitting back waiting. Even finding the right people entailed much effort. Such people would have to be persuaded and motivated and led through the process of innovation. Suzuki had the vision and the passion, and only he could make it all come together.

In the end, Suzuki was able to meet or exceed all of his no-compromise objectives, and the rest is history. At the time of the *Lexus* launch in 1989, Mercedes Benz's three models (300E, 420SE, 560SEL) had no rival in the U.S. market. But *Lexus*, with only a single model was able to sell 2.7 times the number of all three of those well-established Mercedes combined—in a single year. In addition to creating a new luxury division for Toyota, the *Lexus* project injected a new spirit of innovation into Toyota's engineering. It broke the behavioral mold of a global powerhouse with clearly delineated product families, and engineers who had only known the risk-averse Toyota, were suddenly engaged in a bold new challenging project. This renewed spirit would carry over into an entirely new project with new objectives and challenges and would permanently reinvent Toyota's PD process. That project was the *Prius*.

Prius: A New Chief Engineer and New Engineering Process for a Twenty-first Century Car

Like the *Lexus*, the *Prius* was the brainchild of the highest-level executives at Toyota. In the early 1990s, at the peak of the Japanese bubble economy, Toyota's business was booming. Toyota Chairman, Eiji Toyoda, knew this could not last forever. At a Toyota board meeting he asked his colleagues, "Should we continue building cars as we have been doing? Can we survive in the twenty-first century with the type of R&D that we are doing?" In September 1993, Yoshiro Kimbara, then executive VP of R&D, following Eiji Toyoda's lead, founded a project called Global 21 (G21). The project committee was tasked with researching a new car for the twenty-first century—a fuel-efficient, small-sized car with a spacious cabin. The fuel economy target was set at 1.5 times the efficiency of existing small cars like *Corolla*. The vision represented a major design challenge, and no one at the time was thinking hybrid.

Once the G21 had broadly defined the concept for the twenty-first century car, it was ready to name the chief engineer to head up the PD program. In July 1994, Toyota did something that was atypical; they appointed a chief engineer who was not on a career track to become one: Takeshi Uchiyamada. Mr. Uchiyamada had been attending some of the early G21 meetings mainly because he had been the chief architect of Toyota's reorganization of product development into vehicle centers (see Chapter 8).

His task was to figure out where the G21 would eventually fit into the structure. He was surprised and nervous at being named CE. He had no experience in visiting suppliers or checking the production line to solve problems. One of the personifications of a CE is that they "know" everything—from the smallest part, like a bolt, to knowing what the customer wants. Uchiyamada did not and felt unqualified to accept this position. Hindsight reveals, however, that he had three characteristics that made him ideal for this project:

1. *Research Background:* Uchiyamada's background was in test engineering. He was from the research side of the house, which gave him access to an extensive research network to develop the new technology for the twenty-first century car.
2. *Organizational Skills:* He knew Toyota's organization and how to get the resources needed for this unusual project. He also had a talent for creating new forms of organization to move programs forward aggressively. This meant innovatively changing the old method of developing cars, a skill that made him one of the key architects of the largest reorganization of product development in Toyota history (see Chapter 8).
5 *Outside the Box CE:* As a neophyte who had not been "groomed" for the typical CE position, Uchiyamada brought a fresh perspective to product and process.

By intentionally selecting a nonexpert chief engineer, the Toyota executives chose a CE that had to "invent" a new method for developing cars—one of the challenges put forth by the G21.

The first thing Uchiyamada did was to surround himself with a cross-functional team of experts on whom he was to rely to a far greater extent than the traditional CE. This team and leader reliance dynamic brought about a new organizational design perspective, the *obeya* (big room) system of development, which is now a Toyota standard method for developing vehicles. Unlike previous CEs who traveled about meeting with people as needed to coordinate the program, Uchiyamada collocated with a group of experts in the *obeya*, outside the fray of normal day-to-day affairs, to review the progress of the program and discuss key decisions.

As for guiding the development of the *Prius*, Uchiyamada turned out to be a disciplined leader who kept the project on track with respect to timing and cost and *with a no-compromise attitude* to achieving breakthrough targets and desired features. He constantly questioned the purpose of the project and the design concept that would give customers something special. For example, in the early stages of concept development, the team became bogged down in discussing technical details of powertrain technology. Mr. Uchiyamada called the team together and said:

> Let's stop focusing on hardware. We engineers tend to focus on hardware. However, what we need to do with this car is to focus on the "soft" aspects, not the hardware. Let's forget everything about hardware and review from the beginning the concept of the car that we are trying to build from the ground up (Itazaki, p. 46).

Uchiyamada then *got his hands dirty*, leading a brainstorming session of key concepts describing vehicle characteristics of the twenty-first century car. Several days later, the team identified two ideas they believed would define and drive all subsequent development of a "small, fuel-efficient car": 1) *natural resources* and 2) *environment*. The phrase "small, fuel-efficient car" became the G21 goal.

Uchiyamada took another unconventional approach to achieve the G21 goal of a spacious interior. Unlike traditional CEs, Uchiyamada was unfamiliar with the standard dimensions of vehicles, so he turned to his team and asked them to study 30 existing Toyota models. In each case, the team noted differences from the dimensions that the CE team had designed for the G21 project and then asked designers of the 30 existing models why they had come up with their dimensions. The answer was "they had always done it that way." It was only then that the team realized that the reasoning behind its G21 dimensions was sound.

Uchiyamada's approach of having deep discussions and considering many alternative designs early on illustrates the set-based approach (discussed in Chapter 4) that characterized all phases of *Prius* development. The use of the set-based approach by Uchiyamada is even more striking given the intense time pressures the program was under throughout the development process. Figure 7-2 summarizes the time line for the program. As the figure shows, management continually challenged Uchiyamada to do more and more in less and less time. Despite this pressure, Uchiyamada continued to follow the lean PD process of thoroughly considering alternatives before moving on.

As the G21 program progressed, executives escalated pressure on the CE to make a hybrid rather than a conventional fuel-efficient car while, at the same time, pushing for a quick launch. As previously noted, the Toyota

Date	Technical	Organizational Issues
1990	Toyota PD becoming routine, stagnant	Japanese bubble economy peak: Eiji Toyoda preaches crisis mentality
Sep-93	Phase I: Develop car for 21st century	G21 founded by Uoshiro Kimbara (Exec. VP of R&D) with Eiji Toyoda endorsement
Dec-93	G21 concept presented to board (1.5x fuel economy)	G21 study group disbands
Jan-94	Phase II: Create G21 blueprint (detailed concept)	"Permanent project team" for G21 formed
Jul-94	Final report of detailed concept	Team disbands; members return to develop components for G21 in functional homes
Jul-94	Phase III: Actual development for G21 begins	Form development team led by Uchiyamada as chief engineer
Sep-94	Request to present G21 as concept for Tokyo Auto Show (plan on conventional powertrain)	Team must speed development; named *Prius* ("prior" to 21st century)
Nov-94	Mr. Wada (senior technical officer) requests showcase hybrid (double fuel economy)	Secret intention is to push team toward hybrid technology

Figure 7-2. *Prius* Development History

Date	Technical	Organizational Issues
May-95	Present selected hybrid engine type to G21 team	BR-VF does exhaustive search of 80 hybrid engine types and settles on one as "de-facto standard."
Jun-95	G21 becomes official development project	All board member meeting; decide on people, dollars, timing
Aug-95	Plan developed (1st prototype in year; thorough research year 2; production model year 3; SOP end of 1998 at earliest)	Okuda makes "impossible" request—Launch by 12/97
Oct-95	First public presentation of Prius at Tokyo Auto Show (showcase product for Toyota)	By show have decided to pursue hybrid for Prius
Feb-96	First round of sketches for vehicle style (>20 to 5 designs selected)	Used competition among design centers
Jul-96	Final styling review—select innovative Calty design	By board, General Managers, and broad panel; Line off (first vehicle) set for 12/97 (17 months out) and official board review set for 9/96
Dec-97	Launch Prius	

Figure 7-2. *continued*

chief engineer is fiercely independent and intentionally given a great deal of autonomy to run an entire program. While this is true of the day-to-day operation of the program, higher-level executives certainly have a hand in influencing the overall targets for the program, including performance features, timing, and cost targets. In September 1994, the CE team met with Executive VP Wada and Managing Director Shiomi. During this meeting, hybrid technology was discussed but no conclusions for pursuing it were reached. In connection with continuing the development of the G21 project, the G21 group and CE team were asked to present Toyota's small, fuel-efficient car concept for the Tokyo Auto Show scheduled for October 1995. This meant they had a year to develop what would become the showcase product of the auto show. However, this turned out to be the least of their challenges. In November 1994, Mr. Wada casually told the CE team: "By the way, your group is also working on the new concept car for the Motor Show, right? We recently have decided to develop that concept as a hybrid vehicle. That way, it would be easy to explain its fuel economy." (Itazaki, p. 69)

At another meeting with Wada and Shiomi around the same time, it was concluded that the team needed to double the current fuel economy; a 50 percent improvement was deemed far too little for a twenty-first century car. Current engine technology could not achieve this, and Uchiyamada protested as such, but the reply was subtly definite: "Since you are already developing a hybrid vehicle for the Motor Show, there is no reason not to use a hybrid for the production model." (Itazaki, p. 72)

As is often the approach at Toyota, these two executives were not issuing explicit orders to create a hybrid. They were perhaps not so subtly suggesting to the team that this was a natural conclusion. A genuine twenty-first century car had to have breakthrough fuel economy, and a hybrid was the only practical alternative. Uchiyamada took up the challenge and got one important concession—the right to select the finest Toyota engineers available to work on the hybrid engine system.

In less than a year, the new hybrid team developed a new working hybrid system and concept vehicle for the auto show. Work continued on the G21 and in June 1995, the G21 became an official development project with a budget and schedule. Uchiyamada and the team decided to really stretch themselves on timing and committed themselves to a launch date of December 1998 with a slight cushion that would allow the launch to be rescheduled for early 1999. This gave them a year to develop the first complete prototype, a second year to refine the prototype, and a third year to finalize the production version and prepare for manufacturing. Considering that this was new technology with a new production line, this time frame was extremely tight.

The timing pressure intensified even more in August 1995, when the new president of Toyota, Hiroshi Okuda, saw the importance of the G21 project and told Wada the stretch goal had to be earlier because this was the car that could change the course of Toyota's future (Itazaki, p. 115) Uchiyamada was shocked. He was working on a car that could change the course of Toyota's future, but his launch date had been moved up to December 1997. An entire year of the projected time frame had vanished.

The *Prius* was launched in October 1997, two months ahead of the new target date, and potential customers immediately bombarded Toyota with inquiries. At the end of March 1999, sales had reached 22,000 units. By early 2003, this number had risen to 120,000. Worldwide sales since then have continued to grow and the *Prius* continues to be a Toyota success story but for reasons that are not limited to its hybrid engine.

Certainly, the hybrid engine had caused a great stir in the marketplace. But perhaps, more importantly, and for the future of the company, the CE team working on the *Prius* had also made some important and fundamental innovations in the product development process; innovations that are now used in all vehicle development programs at Toyota and are helping to achieve the next target of regular 12-month programs. Gauged by this measure, the returns on the *Prius* project are astronomical and the investment almost trivial.

The CE Leadership Model

In the lean PD system, the standard operating procedure is to give managers more responsibility than they have the authority to carry out. Toyota's CE system epitomizes this. At any one point, there are thousands of Toyota associates working on a program, but the CE has a team of only six to ten people who formally report to him. John Shook, the first American to become a Toyota manager in Japan, was surprised to experience this system and came to describe it as "responsibility without authority." As discussed later in this work (see Chapter 8), Toyota has adapted a matrix organization structure to its CE system. The functional groups, like body engineering and chassis engineering, are technical specialty groups with their own general managers. The general managers supervise the engineers and decide which projects they are assigned to, conduct their performance evaluations, and determine promotions. The chief engineer controls the vehicle program and is responsible for the results but depends on all of the functional groups to supply the people and get the work done. The Toyota Way culture binds the entire enterprise, promoting the common objectives of satisfying customers and making the company successful.

All engineering managers have the responsibility of coordinating a group of people so that their activities align to complete a project. But as Figure 7-3 illustrates, there is a marked difference between how a traditional chief engineer leadership role and a lean, CE type of leadership role handles this responsibility.

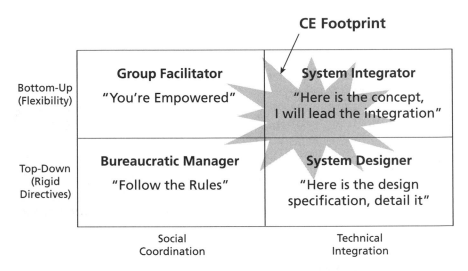

Figure 7-3. Leadership Model: Types of CE Leaders

In the figure, combinations of leadership skills show two dimensions of engineering leadership. In one dimension, an engineering leader focuses primarily on social coordination or technical integration of people and their activities. The second dimension distinguishes top-down leaders who dictate to others what to do versus bottom-up leaders who draw out the expertise of others. When you put these together, you see four types of leaders.

1. *Bureaucratic manager:* This manager coordinates people top down and does not draw on his or her own engineering expertise, relying instead on standards and target schedules and task delegation.

This type of leader follows the engineering standards and rules, meets deadlines and rules by Gantt chart and budget, and compels projects efficiently. There is little flexibility in adapting the timeline or the technical vision beyond routine work, and the bureaucratic manager cannot make creative leaps that will move the project beyond the initial vision and the capabilities of individual engineers. Undoubtedly, to rise to the leadership position, this individual must have had strong engineering skills at some point; in the managerial position; however, he or she employs these skills only minimally. The person can bring in a project on time and on budget but is unlikely to be a great engineer. He has become a project manager, not an engineer or leader.

2. *System designer:* This engineering manager has exceptional technical skills and is passionate about making the product fit a vision of technical excellence, with parts of the system working together to achieve the design objectives. This type of leader is a creative thinker and excellent systems engineer, however, he or she is not terribly skilled at managing people or having the patience to coordinate, teach, or listen to them. This manager has a top-down style for making key technical decisions and uses subordinates to do the detailed, routine work. Henry Ford, in his early days, fits this model. There are limits to the flexibility of the project because many people are doing the detailed work, all orchestrated from the top, and this works only if everyone is following clearly specified instructions. Changes will reverberate through the organization, and the teams are incapable of thinking for themselves without direction from the top.

3. *Group facilitator:* This engineering manager is a people person who has developed leadership skills and is able to take a group of individuals and facilitate their working together as a team. This type of leader is not necessarily a great engineer and may even find detailed technical work boring. Instead, he or she likes to communicate, facilitate, and be a catalyst that moves a talented group of technical professionals toward a common goal. This manager is a flexible thinker and the group can work autonomously to organize and reorganize itself. Terms like "rugby-style management" capture this leadership style, which is characterized by spontaneous adjustment as the game progresses. The weakness in this approach is the lack of a strong technical vision from the

top. This can lead to engineering details falling through the cracks, which can cause time lines to extend outward and weaken technical integration.

4. *System Integrator:* A system integrator is strong technically and uses a bottom-up process to bring out the best ideas from team members. This type of leader has a strong vision for the product and orchestrates the technical integration of the project; he or she also facilitates a dynamic team process with a great deal of flexibility. Toyota CEs best fit this model.

The CE system is designed to give the CE a small number of people to manage administratively, freeing the CE to lead the project by focusing on technical vision and horizontal cross-functional group facilitation. Both Uchiyamada and Suzuki exemplify this leadership style. Each developed strong visions for the product and sought out the right people and resources at the right time. They also excelled at "top-down" management, adhering to strict targets for timing, cost, weight, and fuel efficiency, while at the same time motivating their teams to accomplish remarkable things technically. Note that Figure 7-3 shows that the footprint of the CE cuts across all four types of leaders; residing primarily in the "system integrator" quadrant, he or she also at times puts on the hat of the other three.

NAC Product Development Manager: From Chief Engineer to Bureaucrat

North American Car Company started as a small entrepreneurial venture and, in fact, its early engineers had a lot in common with the chief engineers of Toyota. Leaders of the company loved cars, grew up tinkering with machines, and were more inventors than technicians or managers. They provided a creative genius needed to push the envelope of technology to higher and higher levels. Under their leadership, automobiles became more sophisticated and costs were dramatically reduced. Because NAC was a small company, these "chief engineers" of old were unencumbered by a lot of bureaucracy. In fact, they worked closely with the company's chief owner, who shared their excitement about cars and technology. This gave them a great deal of authority, but their power also came from having strong technical expertise. At that time, the company was quite autocratic; engineering leaders made technical decisions and expected their orders to be followed. They were the architects of the system, and others in specialized functional groups assisted them in handling the detailed tests and design work.

As NAC grew into a global powerhouse, the vehicles became far more varied and complex, and the organizational structure mirrored this. Departments proliferated. Functional "chimneys" were formed to lead every detail of every vehicle—chassis, powertrain, car bodies, electrical systems, electronics, die engineering, engineering analysis, window regulators, emissions control, and on and on. Each functional group had several layers of managers who ultimately reported to the vice president of Engineering. This person was so high up on the organizational chart, there was no bandwidth to keep track of the myriad of work details throughout the company. The overall coordination of the program belonged to specialized program management groups; the technical integration was the responsibility of specialized systems engineering groups.

Lost in this complex bureaucratic organization were the customers. Of course, the company could point to specialized sales and marketing departments responsible for identifying what customers wanted, but by the time that information found its way to disparate technical groups, these customer preferences were distorted beyond recognition. In the second half of the 1990s, NAC moved to a matrix organization in an attempt to compensate for the heavy emphasis on functional departments making separate decisions with weak coordination with other departments. The result was another bureaucratic organization of program managers attempting to exert influence over recalcitrant functional departments. Clearly, the functional departments were winning the power struggle.

Thus, at NAC, bureaucratic managers who learned how to play the game led product development, replacing the powerful and creative chief engineers of old. These managers were administrators rather than system designers. Instead of leading a technical process, they issued orders, set policies, defined objectives for programs (e.g., timing, cost, features), and used rewards and coercive power to whip the organization into shape. At the same time, individual functional departments suboptimized and attempted to grow their functions and power base from the perspective of their own objectives.

Group Facilitation at Chrysler

Durward Sobek's (1997) research comparing Toyota's product development approach to Chrysler's is an interesting case in which Chrysler's original platform team structure was presented as an example of "group facilitation." Like NAC, Chrysler favored the bureaucratic management system up through the early 1990s, and this nearly drove the company out of business. Chrysler was unable to develop new vehicles rapidly enough to get the market share needed to sustain the company. In a bold move, Lee Iacocca and his vice president of engineering, Robert Lutz, reengineered the entire product development organization around platform teams and even built a multibillion dollar R&D center in Auburn Hills, Michigan, to collocate cross-functional platform teams. Chrysler eliminated much of the functional organization that was the backbone of companies like NAC and that continues to be a major strength of Toyota. Once this change was implemented, the functional experts—body engineers, chassis engineers, electrical engineers, component engineers, and even some manufacturing engineers—all reported directly to the head of the platform team. In fact, Glenn Gardner, the first platform team general manager insisted that he either owned these people 100 percent or not at all. This product focus led to great coordination across engineers toward a common goal. Gardner and later platform general managers facilitated groups toward consensus, and product development lead times shrank significantly from 48 months or more to 33 months on the first program and were reduced even more on later programs like *Neon* and the minivan.

In retrospect, it seems that this new approach led to breakthrough projects like completely new, large cars (*Concorde, LHS, Intrepid*), the *Neon*, very new and different minivans, the new *Jeep Grand Cherokee*, and the *P.T. Cruiser*. These fresh new products were launched in record time for Chrysler and saved the company. Moreover, Chrysler boasted the lowest cost per vehicle in the industry and the largest profit per vehicle, making even Toyota look carefully at Chrysler as a potential competitive threat.

But within this apparently successful model, Chrysler engineers spent morning until night in meetings and seemed to focus more on developing group consensus than on engaging in detailed engineering work. Product quality and durability never achieved Toyota levels, and the lead time for developing products never caught up with Toyota lead time. There was also some difficulty in maintaining the deep technical expertise of engineers (who did not work with other specialists in the other platforms), and design standards across platforms did not develop or improve. This, ultimately, led to a creeping increase in product cost. It seems that there were limits to a pure group facilitation approach and to eliminating the strengths of the functional model. In contrast, Toyota's system based on the "system integrator" role has gone farther than Chrysler's and has stood the test of time. It is not surprising that in recent years, the Chrysler group of Daimler-Chrysler has been working to adopt a version of the CE system of Toyota.

Toyota CE System: Avoiding Compromises that Lead to Bureaucracy

Toyota seems to be able to break time-honored organizational principles and avoid compromises. Through the CE system, the company has reaped benefits that derive from its cross-functional product-focused organization, which is peopled by functional experts. It also benefits from top-down management that meets strict timelines and targets and from the flexibility and creativity of bottom-up style management. The Toyota CE shares the technical excellence and passion of the early system designers of NAC. Like those early system designers, Toyota's chief engineers have the ear of top management and are relatively unencumbered by bureaucracy or onerous administrative responsibility. By design, the engineering organization is housed in functional groups, and chief engineers do not have to concern themselves with a great deal of administrative work to manage those engineers and worry about human resource issues.

But Toyota's chief engineer differs from the early system designers in important ways. Toyota is a huge, multinational bureaucracy that cannot function exactly like early small car companies functioned. Toyota's culture is based on consensus management (to a degree) and not blatant autocracy. The size and complexity of the organization and the vehicles it produces prohibits the chief engineer from making all key decisions alone and barking orders. Furthermore, engineers do not formally report to the chief engineer. Thus, the CE must combine his or her recognized technical superiority with strong leadership skills to mobilize the organization and must guide a process of developing consensus across functions. One thing that enables the CE to accomplish this is the concept paper; another is the CE's role as system designer and overall architect of the vehicle. In addition, the chief engineer has the final word on technical decisions (subject to executive approval on major issues). The system works because Toyota has a culture focused on the customer, and functional engineering groups recognize that they exist to *support Toyota's customers* and that the CE is the *voice of the customer*.

Key decisions, mentoring, lobbying for resources, building a shared vision, pushing the product to higher levels, and achieving quality, safety, cost, and timing targets all start with the chief engineer. This makes the CE system stand out as a pivotal part of Toyota's PD system. As the analysis of Toyota's people systems continues, it will become clear that the CE system works because of the broad expertise and organizational alignment that has evolved at Toyota. In connection with this, the following chapter examines LPDS Principle 6 and Toyota's Matrix System.

LPDS Basics for Principle Five

Develop a chief engineer system to lead the development from start to finish

The LPDS is led by an exceptional chief engineer (CE) with the skills to lead system integration, both in the product and integrating the people working on the program. The CE is different from the traditional project manager in several respects. First, the CE does not manage the engineers working on the project, with the exception of a small group of assistants. The CE leads through personal influence, technical know how, and authority over product decisions. Second, the CE represents the voice of the customer and is responsible for the success of the vehicle program from concept to sales. Third, the CE focuses more attention on decisions about systems integration than on personnel decisions and project administration. If a CE is acting only as a project manager, then your company does not really have a CE role.

CHAPTER 8

Organize to Balance Functional Expertise and Cross-Functional Integration

One result of the Prius *development project is something called* obeya—*big room—where the chief engineer gathers the team of people responsible for that project. That is where simultaneous engineering can be even more effectively implemented by all the key people coming together in this area.*

TAKESHI UCHIYAMADA, first chief engineer of the *Prius* hybrid

One Best Organizational Structure?

The problem of attaining the "best" organization structure for product development is finding a balance between trade-offs. Various structures promoting product-focused organizations have replaced the traditional functional organization structures that were most common at the turn of the twentieth century. Now companies see the functional organization as bad and the product-focused organization as good. This seems clear-cut, but the lean PD system uses neither . . . or both.

A functional organization groups like specialties and like-minded professionals into departments. It segregates all of the mechanical engineers into a set of cubicles where they share war stories and practices. All of the electrical engineers occupy a different set of cubicles and gossip about the boring work mechanical engineers do. The manufacturing engineers go to offices on or near the shopfloor where "real" work is being done and criticize the head-in-the-clouds theoretical work of the electrical and mechanical engineers. And so on. At one time, however, the functional organization had certain advantages. These existed because functional specialists:

- could talk to each other efficiently in a specialized language.
- shared the latest technologies and methods with each other, increasing their technical depth.
- attended the same professional conferences, read the same journals, and continued to learn long after leaving school.

- could standardize their approaches and the actual technologies they used in the product, saving cost and sharing solutions to problems.
- could be efficiently deployed as needed to projects (whoever was available at the time could go work on the project)—thus fully utilizing all of the engineering resources.

But there was one big problem with this organizational approach—*functional specialists tended to bond and become more wedded to their function and profession than to the company and its products and customers.* Their measure of success was how well the functional department performed and how big a budget it garnered. They believed their profession could be the savior of the company; if they ran things, the company would be successful beyond belief. As a result, no function did a particularly good job of working in coordination with the other functions. Today, these isolated functions are often denoted by derogatory terms like *functional silos* or *chimneys*.

Shortcomings of a Product Organization

One alternative to the functional organization is the product-focused organization or *product organization*. This organization creates cross-functional teams that focus on a project or product and sets clear goals and objectives for the new PD program, holding the team responsible. It dedicates representatives from all of the needed functions to the program to develop the product and process. And, if possible, it collocates teams so that they will communicate constantly about the product and the customer. This is sometimes called *concurrent engineering* because it is designed to develop product and process concurrently rather than serially (Fleischer and Liker, 1997). Having a product focus breaks down barriers between the "silos" and aligns the company as a whole to care about what is really important—satisfying customers so they will buy more products and make the company profitable. The benefits of the product organization are also clear-cut because they allow companies to:

- align different functions around common goals and objectives that are needed for creating products that satisfy customers.
- communicate and coordinate effectively to reduce lead time.
- make well-informed product and process decisions from multiple perspectives to increase product quality.

- create self-managing teams to be flexible and adaptable to changes in the environment.

But the product organizational approach also poses problems, which can be exemplified by Chrysler's adoption of a *platform team organization*. In this structure, Chrysler categorized vehicles into platforms—large car, small car, minivan, etc., and collocated all the engineers from all the needed functions to engineer a complete vehicle into one floor or a new technical center. All the engineers reported to a "general manager" whose responsibilities were similar to those of Toyota's CE. However, as Sobek (1997) observed, there were associated costs and problems with this restructuring. For instance, Chrysler engineers spent an enormous amount of time coordinating their work and sitting in meetings while some of the standardization within functions (e.g., sharing common parts) got lost. Soon, each platform became its own "chimney," only in this case, it was a product chimney rather than the traditional functional chimney. This led to an inefficient utilization of resources across platforms, further fixing engineers within their respective platforms. Furthermore, even though workloads swing up and down over the life of a program, the general manager wanted to maintain the same number of engineers because "sharing an engineer" with another platform could mean losing that engineer.

Over time, a kind of empire building mentality evolved, and many of the benefits of Chryslers' platform team structure dissipated as people pursued the territorial benefits of their own "product chimneys" (e.g., more engineers, bigger budgets, greater prestige). Top management saw this trap and developed a solution, forming "technology clubs" across platforms to bring together functional specialists to share technical information and develop standards for components. Nevertheless, the technical clubs tended to play second fiddle to the incessant daily demands of the programs—particularly as the platform general managers had authority over the engineers. As G. Glenn Gardner, the first platform general manager, put it, "If I do not own all of a person I do not own any of that person. I need full-time engineers on the platform."

Toyota seldom settles anything with compromises and seeing a choice between a functional organization (with its functional expertise and the efficiencies that come with sharing people across programs) or the product organization that integrates systems across functions, Toyota's view will be, "I want them both." The secret to Toyota's success is combining a strong functional organization based on deep specialization with the CE system as the other leg of the matrix. Using this matrix organizational structure, Toyota manages to have it both ways.

Strengths and Weaknesses of the Matrix Organization to Manage the PD Process

Since the 1960s, many organizations have developed and adopted various versions of the matrix organization, with mixed results. It is in some ways the best of both worlds and in other ways the worst of both worlds. It provides:

- an excellent balance of functional expertise and cross-functional integration.
- the technical depth and efficiency of the functional organization with the customer focus of the product organization.
- flexibility in assigning resources to programs as well as technical depth in responding with creative solutions to new problems.

NASA pioneered the matrix organization structure in the 1960s when it was first contemplating space exploration.[1] It was a solution that satisfied the need for deep functional expertise—to get all the individual vehicle systems exactly right, using the latest technologies, meant that NASA needed every part of the spacecraft to work together as a system. Misalignment could be a matter of life or death for astronauts. NASA also kept the functional organization structure, adding a program management structure to manage the large government projects that are its bread and butter. This meant that engineers reported to at least two people in two different arenas: functional managers within their functional specialty and program managers who were leading each of NASA's space exploration programs.

As NASA's experience demonstrates, the matrix organization has one major flaw: It is confusing! It violates a core principle of personnel management that clearly stipulates that no one should have more than one boss. As the saying goes, "A person with two bosses is like a dog with two heads." This leads to communication and leadership-direction problems—whose orders are to be followed—or engineers taking advantage of the fact there is no clear authority and seeking the approval of the manager most likely to say yes to their demands and needs. This is human

---

1. Toyota developed a form of matrix organization long before NASA.

nature. Moreover, it is an environment in which engineers can play bosses against each other just as children play parents against each other, a situation that can lead to infighting between bosses. Toyota manages to avoid these problems by grooming its CEs to play the program manager role in a matrix in a manner that precludes these problems.

Toyota's Original Matrix Organization: A Long Tradition of Combining Two Structures

How does a lean PD system combine a strong functional organization with a matrix organization and keep the peace between these two structures? The secret is that it uses a combination of the intense customer focus (the core of the Toyota DNA) and the CE system discussed in Chapter 7 to make the program management part of the matrix work in harmony with functions. It was in the 1950s that Toyota merged its unique CE system into its homegrown matrix system.

Figure 8-1 presents a simplified version of Toyota's matrix organization. Originally, each product had one functional department and each

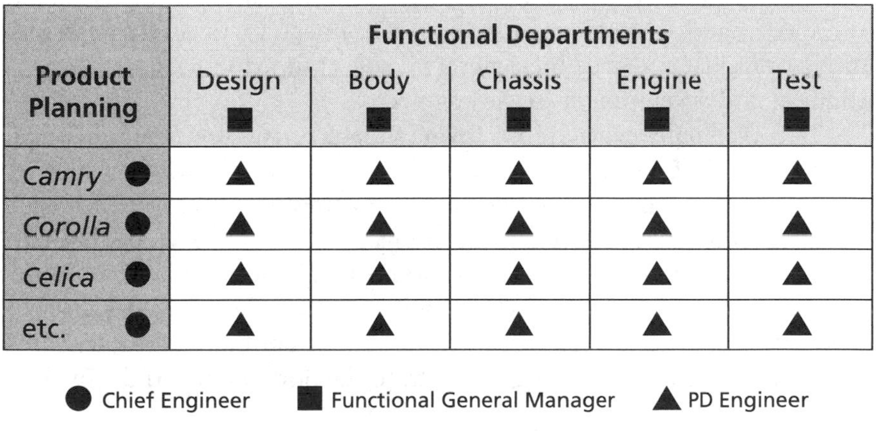

Product Planning	Functional Departments				
	Design ■	Body ■	Chassis ■	Engine ■	Test ■
Camry ●	▲	▲	▲	▲	▲
Corolla ●	▲	▲	▲	▲	▲
Celica ●	▲	▲	▲	▲	▲
etc. ●	▲	▲	▲	▲	▲

● Chief Engineer　　■ Functional General Manager　　▲ PD Engineer

- Most engineers report to functional managers.
- Engineers are assigned to projects as needed.
- Production Engineering is a separate division.
- Chief Engineer has a small staff of assistants.

Figure 8-1. Toyota's Matrix Organization—"It's the Chief Engineer's Car"

functional department reported to a "general manager" who had distinguished himself or herself as an excellent engineer and leader within that particular function. The general manager's job included:

- selecting and developing engineers in the specialty.
- coordinating performance reviews of the engineers reporting to him.
- maintaining checklists of accumulated knowledge for the specialty.
- keeping the specialty state of the art.
- ensuring technical coordination, such as common parts across vehicles.
- working with suppliers of components related to the specialty (e.g., resident engineers who come from the specialty suppliers).
- assigning the engineers to projects run by chief engineers.

Thus, the general manager had the responsibilities of a traditional manager as well as a technical expertise in a given specialty and provided administrative management, developmental leadership, and technical mentoring. What the general manager was not responsible for was the development of a vehicle. That was the sole domain of the chief engineer, who likewise was not responsible for the administration or management of the engineers. This structure gave the CE time to focus on the customer and the product and gave the general managers time to focus on the management and development of the engineers.

This eventually evolved into Toyota's matrix structure: The functional managers make up one side of the matrix and the CEs make up the other side. Normally, a matrix is viewed as a reporting structure in which each engineer reports to a functional boss and a program management boss. At Toyota, most engineers do not report to the CE. Instead, there is a "dotted line" reporting relationship: When engineers work on a CE's design program, they are mainly reporting to the functional boss but dotted line reporting to the CE. So what gives teeth to the dotted line? And why does this work at Toyota?

To answer these questions, it is necessary to examine why so many companies struggle with the matrix organization. As noted above, the engineer in most matrix organizations has two bosses—the general manager and the program manager. Because the general manager handles his or her performance review, the engineer is more inclined to favor the general manager. The Toyota matrix system would seem to almost guarantee this bias by having the engineer report administratively to the functional boss. Why would the working engineer pay any attention at all to the chief engineer and the specific vehicle program? There are five reasons that Toyota can successfully integrate the functions and programs:

1. *Customer is first.* From the moment an engineer joins Toyota, he or she is never allowed to forget this principle. In Japan, new Toyota engineers are even assigned to sell cars door to door, which drills into them the customer-is-first philosophy—an engineer exists to serve the customer, not the functional boss. Simultaneously, the engineer learns that the CE represents the customer. The chief engineer controls the vehicle program by representing the *voice of the customer*, so the top priority is for the engineers to listen to the chief engineer. This cultural philosophy, and the behavior that reflects it, is unusually strong in Toyota and essential to a lean PD process.

2. *Chief engineer is revered.* As discussed in Chapter 7, every engineer recognizes the engineering skill, leadership skill, and dedication it takes to become a chief engineer. This merits a high level of respect and compels every engineer to support the chief engineer.

3. *Chief engineer has the power of executive sponsorship.* It is the chief engineer's car and the chief engineer can tap any executive any time for help. The CE may not have formal authority, but he or she has plenty of access to a pool of executive authority.

4. *General managers understand the importance of serving customers and cross-functional cooperation.* When conducting performance reviews, the general manager takes very seriously the need for soliciting feedback from other functions and from each engineer's CE. Thus, the engineer sees every senior official as a boss who can influence his or her performance review.

5. *Junior employees respect senior employees.* The Japanese culture reinforces the respect given to senior employees with more experience. Junior employees naturally defer to senior leadership, which reinforces learning and clarity of authority.

A Fundamental Change to Toyota's Matrix Organization

Toyota's matrix organization worked just fine for decades, but as Toyota grew and cars became more complex, the functional specialties proliferated. In 1976, the CE had to coordinate people from 23 departments in

6 divisions; by 1991, this had doubled to 48 departments in 12 divisions (Cusumano and Nobeoka, 1998). As the number of programs grew, functional managers had a hard time managing the details of so many programs. Across the board, the system became too complex to manage.

In 1992, Toyota did something unusual. After decades of fine-tuning one type of matrix organizational structure, the company fundamentally reorganized it around three *vehicle development centers*, each of which handled a different product family of vehicle platforms: rear-wheel drive, front-wheel drive, and utility vehicles/van. In 1993, Toyota added a fourth center to develop components and systems that cut across vehicle platforms. This fourth center also housed most research and advanced development, along with some general electrical engineering and engine development work. In 1993, about 12,000 people worked on product development. With the reorganization, each of the platform centers housed under 1,900 people (with the largest group located in Center IV) working on about five new vehicle programs simultaneously.

The result was smaller vehicle centers that considerably reduced the coordination requirements for chief engineers and, in fact, allowed for much greater coordination across the programs within a vehicle center. To simplify the coordination requirements further, Toyota simplified the functional departments. For example, two different divisions for body engineering—interior engineering and exterior body structures—became one body-engineering department within each of vehicle development Centers I through III. Similarly, Toyota merged two chassis engineering specialties into one, reducing the number of handoffs across functional departments. Now, with three smaller organizations, the functional general managers had fewer programs to support and could handle the added specialties within their respective domains.

Each vehicle center also has its own planning divisions supported by almost 200 planning people—about 10 percent of the employees in each center. These planning divisions include the CEs and their staffs, who managed the programs, and planners, who kept track of costs and timing on every detail of each program. Even today, Toyota's extensive planning means that costs and timing relative to targets can be reported daily, which enables Toyota to execute programs smoothly.

The vehicle center structure changed once more with the rise of *Lexus* and the desire to separate this luxury car division from the rest of the products. Figure 8-2 depicts the resulting matrix organization (earlier version was in Cusumano and Nobeoka, 1998). Each center looks much like

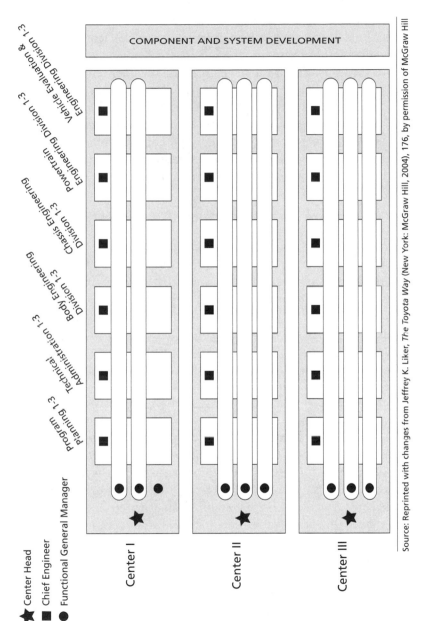

COMPONENT AND SYSTEM DEVELOPMENT

Center I

Center II

Center III

Engineering Division 1-3
Vehicle Evaluation &
Powertrain
Engineering Division 1-3
Chassis Engineering
Division 1-3
Body Engineering
Division 1-3
Technical
Administration 1-3
Program
Planning 1-3

★ Center Head
■ Chief Engineer
● Functional General Manager

Figure 8-2. Toyota's Vehicle Development Center Structure

Source: Reprinted with changes from Jeffrey K. Liker, *The Toyota Way* (New York: McGraw Hill, 2004), 176, by permission of McGraw Hill

a smaller version of the original matrix organization. Most of the rear-wheel drive vehicles are in Center I, and most of the front-wheel drive vehicles are in Center II. The trucks and chassis-based sport utility vehicles that had been in Center II were moved to Center I. A few front-wheel drive vehicles are in Center I (e.g., the U.S.-based *Avalon, Camry, Sienna, Solara* are in Center I). The third center is devoted exclusively to all *Lexus* vehicles. *Lexus* simply has higher standards for everything, engineering and manufacturing, and Toyota decided to keep it separate from other products so it does not become a standard Toyota with a *Lexus* nameplate. Note that the centralization of component and core system development concentrates these developments as a shared resource to the development centers so that they can be commonized across programs.

Chrysler's Platform Team Structure: A Contrast to Vehicle Development Centers

It has become quite common for automotive companies to organize their PD program around vehicle platforms, but most do so without the CE role. Sobek (1997) documented Chrysler's *platform team organization*, which provides an excellent contrast to Toyota's *vehicle development centers* and the role of the CE in the Toyota matrix organization structure.

The structure for Chrysler's platform team organization was originally set up in 1989, about three years before Toyota's reorganization. Chrysler's platform team organization was the company's response to problems that had arisen because the functions within its functional organization did not communicate well or work well together. Company executives decided they needed a radical change: to blow up this functional organization and turn it completely into a product organization. At first, Chrysler launched the platform team structure with the large car LH platform, which was the large passenger car front-wheel drive platform (Dodge Intrepid, Chrysler LHS, Chrysler Concorde, etc.). At the time, Vice President Bob Lutz had already made the decision to convert Chrysler's purely functional organization to platform teams across the board, and Chrysler soon created platform teams for its small car, minivan, Jeep, and truck divisions. The rapid change was motivated by one simple reason: Chrysler was on the brink of bankruptcy. It was riding the long wave of the k-car—produced on a single and very old platform—and desperately needed a new product. Survival meant change. In this environment, it was a bold move for Iacocca to benchmark Honda's PD system and then bet the company's future on the platform team organization. The concept was to pour the functional organization into a product organization, using cross-functional teamwork as the driving force behind the reorganization. As mentioned in Chapter 7, Chrysler collocated engineers from across functions in a newly built technical center in Auburn Hills, Michigan, designed to house a whole platform in one part of one floor, so teams could continuously interact.

G. Glenn Gardner headed the LH platform program and became Chrysler's first "Platform General Manager." Earlier in his career, Gardner had, in fact helped shape the concept for the platform teams, leading to the development of a number of successful Chrysler vehicles produced in a "skunkworks" fashion. The teams had been set up, housed offsite, removed from the Chrysler function-based bureaucracy, and given the freedom to focus on the vehicle. The results were outstanding. Unfortunately, Chrysler later disbanded the teams; team members returned to their home functional departments, and all of their knowledge was lost. After some experience working as an executive at Mitsubishi, where he learned about cross-functional teamwork, Gardner returned to Chrysler. Armed with new experience and new knowledge, he had no interest in reviving the skunkwork approach; he wanted an approach that would sustain PD teams from project to project. He insisted that:

- Chrysler supply him with all of the engineering specialists needed to get the car designed.
- the manufacturing process engineers reported to him.
- every core engineer was on the project team 100 percent of the time—not split with other responsibilities (the matrix was not an option).

Gardner also insisted that if he had a well-selected team reporting to him 100 percent of the time, he could do the job with about half the number of people working on traditional Chrysler programs. True to his prediction, the LH team reduced the development time for the LH family of vehicles from 4.5 to 3.5 years and did the job with 741 people instead of the 1,400 people originally forecasted, saving $42 million. The overall product came in under weight, $20 under target cost, and had better fuel economy than targeted. In addition, it was a completely new platform with a complete line of large cars and a new engine. The post-launch results were stunning. The LH platform cars were attractive, well packaged, competitively priced, and started the resurgence of Chrysler. The vehicles that followed in the other platforms were all successful in their own right, and Chrysler became the low-cost producer of vehicles with high profit margin per vehicle.

As Figure 8-3 shows, the platform team organization that Chrysler used was cross-functional and included engineering teams from body, interior systems, chassis, powertrain, electrical/electronics, preprogram, and the vehicle evaluation departments (Sobek, 1997).

- Five platforms: large car, small car, minivan, Jeep, truck.
- Platform lead by cross-functional Platform Direction Team:
  Engineering member = Platform General Manager.
- Team members dedicated and colocated.
- Integrate across functions in platforms through technology clubs.

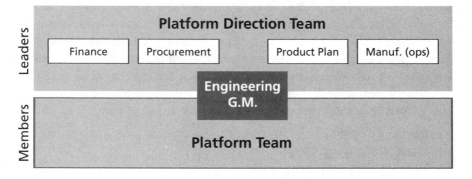

Figure 8-3. Chrysler's Platform System in 1989

An "executive engineer" heading each of these areas reported to the general manager (just as Gardner had insisted), who ran the program with a small staff of program managers similar to Toyota's CE staff. The cross-functional *platform direction team* ran the platform, while the product engineer (the general manager) ran day-to-day operations. The platform direction team included finance, procurement, product planning, and manufacturing operations. The four functional groups each had representatives assigned to the platform direction team who reported in a dotted line to the platform general manager. The role of the platform direction team was to get the nonengineering cross-functional support to focus on the customer and the product and stop territorial disputes that often occur among nonengineering and engineering functions.

Since so much of the emphasis of the platform team structure is to coordinate across different functional specialties, a great deal of time and resources was devoted to communication. In Chrysler's previous platform organization, the platform engineering team did their work within their specialty and then "threw it over the wall" to other groups who challenged technical specifications and then threw the design back over the wall. This sucked up a lot of time and energy and produced inferior designs. The new platform team approach "got all the right functions together in the same room," accelerated the negotiation process, and helped everyone leave the room on the same design page. When compared to its older dysfunctional system, the platform team approach was a highly successful endeavor that revolutionized the PD process at Chrysler. On the other hand, compared to the lean PD system, it had several weaknesses:

1. *Poor resource utilization of engineers.* There is some waste of resources because Chrysler's engineers were committed to a program from start to finish. In the life of a program, different numbers of engineers are needed at different times. Toyota addresses this fact by using the matrix organization to add and remove people from the program as needed.

2. *Poor use of the platform general manager's responsibilities.* The dual role of the Chrysler platform general manager as head of functional general managers and something akin to the CE at Toyota meant this individual spent a good portion of his or her time on administration. This cut into the time spent on the systems integration role, which the CE provides at Toyota.

3. *Poor use of meetings.* Chrysler engineers had to spend a vast amount of time in meetings, often going over administrative things that took time away from working on technical product development. Since they were working in cross-functional teams, they had to listen to reports from all other functions, whether the information was relevant or not. Sobek (1997) calls this communication by immersion in all of the details. Rather than communication by immersion, Toyota engineers do their own CAD work, spending a lot of time at the terminal designing, using a pull system, and seeking the information they need, when they need it, to complete their tasks.

4. *Poor coordination of engineers within their functional specialties.* The Chrysler platform teams focused so heavily on their own platforms,

that engineers made little effort to coordinate their efforts within their own functional specialties. For example, taking the time to standardize parts within a function across vehicles suffered. Chrysler added "technology clubs" so functional specialists, like electronics engineers from different platforms, could meet and discuss common issues, but this often took a backseat as each engineer had to tend to the heavy demands of his or her own vehicle program. In contrast, Toyota's vehicle-development center system takes the time to develop a deeper functional expertise, capturing the learning and applying and/or standardizing it across vehicle platforms.

It is easy to understand why Chrysler did what it did. Because of the severe coordination problems across functional specialties, a radical approach was devised: tearing down the functional organization and replacing it with a completely new PD process that would allow engineers to focus solely on the product. Meetings, morning to night, would force coordination and assign everyone to one hierarchy dedicated to each platform. This worked to a point. However, these efforts did not lead to the vast improvements in lead time that a CE-driven lean PD process delivers. (Toyota's vehicle center system, for example, eventually reduced lead times to 12 to 15 month programs.) By 2003, the Chrysler division of Daimler-Chrysler was looking more closely at Toyota's model. It has since strengthened the functional organization structure and created a CE-like role.

During Chrysler's PD reorganization, Toyota was not standing still. Company executives realized that they needed to revisit their strong functional organization, which, at the time, was being solely driven by the CE system. When Toyota started improving simultaneous engineering—to improve integration of product and production system design—it became clear that the next step of evolution required greater horizontal coordination across the functional organization. This led to additional innovations, such as the obeya room, module development teams, chief production engineers, and a slightly different role for the CE.

Simultaneous Engineering: The *Obeya* Room

In the early 1990s, Eiji Toyoda was concerned that Toyota was becoming fat and stagnant. He wanted to renew the innovative and exciting environment of the early years, and a key part of that strategy was to use the G21

(Global 21) platform—which became *Prius*—for change (for a detailed discussion of this, see Chapter 7). One of the two objectives of that project was to, "develop a new method of developing cars for the twenty-first century." The person chosen to head the G21 project was Uchiyamada, someone who had no experience in product development or as a chief engineer. Uchiyamada had to innovate or fail and his solution was to seek help from others, specifically from individuals who had spent their careers in product development, instead of leading the project alone. The best way to do this was to get technical leaders from each of the core functions together in one room and ask them to help him lead the program. While this may sound similar to Chrysler's platform-direction team approach, there is an important difference. Chrysler designed the platform direction team to solicit broad participation from what are usually support functions, such as purchasing and finance. Uchiyamada wanted technical assistance from core engineering functions to help lead the actual development of the vehicle. Moreover, there was no mistaking that Uchiyamada was still the CE and the final authority on product decisions.

As discussed in Chapter 7, one of Uchiyamada's first innovations was the *obeya*, where once every two days or so the CE participated in a face-to-face meeting with a team of specialists from the various design, evaluation, and manufacturing functional groups. Here, the experts could work with the CE to formulate ideas, address immediate issues, and make on-the-spot decisions. Production engineering also joined in these meetings to discuss and work on issues with the design engineers. Having the teams meet in obeyas served two main purposes—information gathering and information management. Information gathering is mainly the responsibility of functional groups. Information management is about hammering out and communicating the day-to-day decisions in a kind of war room atmosphere, where the CE makes joint decisions with other leaders almost immediately (not in days or weeks). The obeyas were equipped with simple visual management tools laid out densely on the walls, such as graphs of key metrics relative to target to support decision making and schedules with checkpoints so that everyone could easily see every aspect of the program, further forging a common understanding among the teams.

The obeya did not replace Toyota's matrix structure, which had not changed. It was an enhancement that added to the traditional CE approach. Here, the CE would develop the vehicle-concept and then discuss it with the design and planning groups. He would then retreat and write a plan, or concept paper. In addition, while the CE continued to control every aspect of the project, the *obeya* gave the cross-functional team more direct input into decisions as they were being made.

Uchiyamada's second innovation was to have the CE office do all the scheduling. In the traditional CE system, all the teams put schedules together and reported to the CE. In the new approach, the CE used the *obeya* to manage this schedule, identifying different problem areas and setting up task teams. Then the CE assigned a team member to be responsible for coming up with countermeasures for any problem.

A third innovation emerged from Toyota's need to embrace new technologies for the twenty-first century and was simultaneously applied by Uchiyamada to the *Prius* development process. For the first time in Toyota's history, the development team used the Internet and email extensively as a major communication media. Prior to this, Toyota had used information technology conservatively. Uchiyamada came from the research labs and was very comfortable with new technology.

A fourth innovation was to extend the impact of simultaneous engineering and the simultaneous engineer (SE). Toyota had worked on simultaneous engineering for several years before the *Prius* program, but never to the extent that Uchiyamada planned to use it. Because the *Prius* program involved so much new engineering and such a degree of innovative development and because top management was putting intense time pressure on the team, Uchiyamada accelerated the use of the SEs. His plan met with great success and began a new tradition at Toyota. As noted in Chapter 4, there is now at least one SE on each of the several MDTs who act as a program-dedicated representative from production engineering. This SE is a production engineering expert who advises the program MDT on manufacturing aspects of design within his or her specialty (i.e. stamping, welding, etc.) and also serves as the program representative who advises production engineering specialists who will perform the actual manufacturing engineering. In this dual capacity, the production engineer helps to synchronize activities, reduce errors and rework, and build cooperation across divisions. Thus, simultaneous engineering is a vital *horizontal coordinating mechanism*.

Simultaneous Engineering: The Module Development Teams and Chief Production Engineers

At this time, Toyota was also developing another simultaneous engineering initiative, which consisted of two main initiatives: 1) the modular development team (MDT) and 2) a corresponding role to the CE, the chief production engineer.

To some extent, these new initiatives represented a rather strange departure from a system that seemed to be working very well. For many years, in fact, Toyota had enjoyed a reputation for designing products for manufacturability. It was able to maintain this reputation for the following reasons:

1. A prerequisite for becoming a first-rate PD engineer was to spend time in manufacturing to understand the manufacturing environment.

2. Toyota's customer focus mantra made it imperative that PD engineers consider manufacturability technologies and issues.

3. PD engineers had to take production engineers seriously because of the prominence of manufacturing, which makes production engineering influential.

4. Both product engineering and production engineering kept, used, and updated extensive checklists separately, ensuring serious consideration of production in the product development stage.

5. Production engineers attended key reviews along with the PD engineers in the *kentou* (study drawing) stage to provide input and were required to sign off on the design structures document (the K4, which was discussed in Chapter 4).

However, as Toyota continued to streamline its product development process (in some cases eliminating physical prototypes and radically reducing tool-manufacturing times), the old interaction between PD engineers and production engineers proved inadequate. It became clear that Toyota's new, high-velocity product development process required seamless collaboration. Moreover, Toyota wanted to improve its manufacturing efficiency to a level that would allow the company to compete with low labor rates in China. Company leaders realized less labor in manufacturing depended on products designed in optimize labor utilization. In connection with this, Toyota determined that a stronger and more intensive involvement between product engineering and production engineering at a higher level was needed to coordinate the extra complexity and need for speed. It was with this in mind that the company created the module development team (MDT) structure.

Example of Module Development Teams for Body and Production Engineering

Figures 8-4 and 8-5 illustrate MDTs for body and structures engineering. The two examples are from the Toyota Technical Center (TTC) in Ann Arbor, Michigan, and the corresponding production engineering organization in Erlanger, Kentucky, at Toyota Motor Manufacturing North America headquarters.

Product engineers, housed at TTC, were organized by their respective specialties (body engineering, electrical engineering, HVAC etc.) and, in the case of body engineering, further organized by area of the car body— front end, body in white (the car body structure), closures, and the underbody (Figure 8-4). Each of these body subsystems has an engineering team for each vehicle program (*Camry, Avalon,* etc.) led by a senior engineer. From the production engineering perspective, they are organized into comparable groups by function (stamping and structures, paint, final vehicle assembly, etc.) and often further specialized within those functional organizations.

For example, within the stamping and structures organization, engineers are further organized by car body area very much like their colleagues in product engineering (closures, front end, body in white). Each of these areas develops highly skilled, experienced engineers who become SEs and represent their functional specialty on the program within the various cross-functional MDTs. These teams meet regularly from the earliest stages of the *kentou*—before the clay model has even been selected. They work together as an extended team, making decisions jointly. This supplements the process in which PD engineers make decisions based on checklists and then ask production engineers to review these decisions.

Moreover, in North America, Toyota has created the role of chief production engineer (CPE) to have overall responsibility for production preparation and launch for each vehicle program, similar to the CE role in product development. The CPE's role is to lead the overall development of the manufacturing process and coordinate the activities of the SEs across all MDTs. In Toyota's U.S. facilities, an important job for the CPE is to coordinate with Japan (where most of the original production engineering now occurs) and take a leading role in the transfer of the production equipment to the U.S. plant. The creation of the CPE position further demonstrates the challenges of coordinating the development of a complex product in a high-velocity environment.

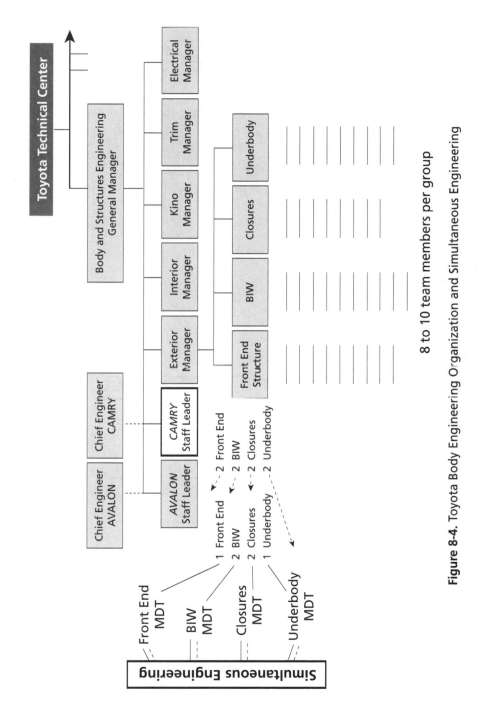

Figure 8-4. Toyota Body Engineering Organization and Simultaneous Engineering

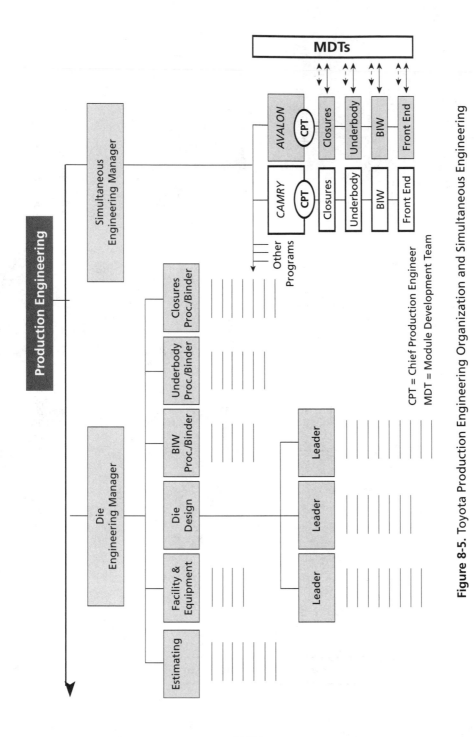

Figure 8-5. Toyota Production Engineering Organization and Simultaneous Engineering

CPT = Chief Production Engineer
MDT = Module Development Team

During the early stages of the *kentou* process (discussed in Chapter 4) production engineers from the SE group study electronically transmitted design information before attending the biweekly modular development team meetings. With production data and updated *senzu* (detailed marked up manufacturing drawings) and checklists in hand, they discuss the analysis of current design proposals. Then the MDT will spend an intensive period analyzing, discussing and codeveloping multiple design alternatives. Because he owns the part from cradle to grave, the SE is motivated to evaluate the quality and productivity implications for manufacturing of the proposed designs.

The MDTs are an invaluable tool for early problem solving, coproduct/process development, and isolating variability. The MDTs are also a critical integration tool in which engineers from different disciplines come together without being collocated and gain valuable insights into the challenges of each other's specialties. As shown in figures 8-4 and 8-5, each production engineer's home base remains within his or her functional specialty; the production engineers also travel to the Toyota Technical Center (TTC) as required. During times of peak activity, the production engineers and others may travel to TTC for a week at a time. Temporary work areas are maintained at TTC for the use of the many MDTs who are working closely with product development at any given time. Toyota MDTs contribute much of the program integration and communication benefits afforded by platform teams while maintaining the superior functional knowledge sharing and personnel development made possible by functional groups. Although only broadly utilized since 1997, MDTs have already had a significant and positive impact on Toyota PD performance.

While the *obeya* and the MDTs evolved as parallel, separate systems, they eventually merged into a unified *Obeya system*. In fact, MDTs are now considered part of the *obeya* process and provide key input into *obeya* for decision making and problem solving. This expanded *obeya* system has aligned the CE with the cross-functional teams, strengthened Toyota's matrix system, and, with the use of the SEs, provided horizontal coordination across divisions, teams, and production. *Obeya* have also evolved to a migratory construct; the main *obeya* is in product development early in the program and moves to the *manufacturing* site in the stage of production preparation for launch. It is important to note that none of this supersedes the matrix structure and the role of the CE and functional organization. That is still the main organization structure and very much intact. *Obeya* and MDTs are additional integrating mechanisms to tie together different functional groups.

Organization as an Evolving Process

Earlier in this book, we mentioned that understanding Toyota's PD system was analogous to removing layers of an onion and, as the last two chapters clearly show, Toyota's approach to organizing product and process development is clearly an evolving and living process. This is part of the power of lean thinking, which emphasizes a learning organization—you are always improving on what you have while maintaining a considerable continuity in the process. Even with the changes, Toyota's PD system is philosophically and fundamentally the same as it has been for decades. The CE system is still vibrant, and the functional organization has been tweaked and streamlined for the twenty-first century. Production engineering is still as big and powerful as ever. However, product development, which had become increasingly complex and unwieldy, has been broken down into smaller, more manageable vehicle-platform chunks. Toyota has, to some extent, reined in the CEs within the vehicle development centers, placing more emphasis on parts commonization. In addition, there has been a deliberate and extensive addition of horizontal-coordinating mechanisms through *obeya* and module development teams and SEs. Even after all these changes, the CE's talent, personality, and perseverance still determine the success of a car.

In following the second people LPDS principle, *Organize to Balance Functional Expertise and Cross-Functional Integration*, your company is laying the foundation for best utilization of your people resources and decision-making responsibilities. The next chapter builds on the discussion of the chief engineer and matrix organization and addresses the third people LPDS principle: *Develop Towering Technical Competence in All Engineers*, which reflects the lean philosophy of selecting, hiring, and developing appropriate talent.

LPDS Basics for Principle Six

Organize to balance functional expertise and cross-functional integration

Deep functional expertise is a requirement for LPDS but so is coordination across functions to stay focused on the customer first. That is why the matrix organization structure is so popular—it provides an opportunity to balance the functional organization and the product organization. But the matrix in reality is often unbalanced. Either the functional organization is dominant, leading to a bunch of technical specialists across functions that do not talk to each other, or the product organization is dominant, and the company loses depth of expertise and standardization of product features. Toyota strikes the balance by beginning with the development of towering technical competence within functions and then layering on this the chief engineer system to keep functional specialists focused on the customer and the vehicle.

CHAPTER 9

Develop Towering Technical Competence in All Engineers

*We develop people and new products simultaneously
using the Toyota way*

UCHI OKAMOTA, former Vice President,
N. A. Body and Structures Engineering

TO EXCEL AT THE TALENT-DRIVEN BUSINESS of product development, a company must have highly-skilled, capable, motivated people. Achieving a lean PD process with a precise and synchronized execution of a leveled flow means that everyone working on a PD program must do his or her job correctly and on time. One weak link can disrupt the precise timing and bring everything to a standstill. To prevent such disruption and ensure success, a company must be willing to make major investments in the process of selecting and developing technical competence in all of its engineers.

In a lean system, people learn best from a combination of direct experience and mentoring. Excellent engineers that fit in with a high-performance PD do not graduate from college ready baked to handle important projects; they are built slowly and from scratch. Toyota has always recognized this truth and has developed rigorous selection and training processes to support it. In connection with this, Toyota engineers travel a career path based on demonstrated competence, a crucial factor in a high-velocity PD process. Speed in product development depends on "professional trust." Engineers in product development are like hockey players or members of a special-forces team—they must be able to trust that others are doing what they should be doing when they should be doing it. In turn, others must be able to trust them. This professional trust has two elements:

1. Integrity. People must have the intent to do what they say
   they will.
2. Competence. They must be capable of doing it.

Professional trust defined by integrity and competence can only evolve over time. It is rooted in rigorous selection and training, and grows between professionals who consistently demonstrate battle-tested, reliable performance.

A Philosophy for Hiring, Developing, and Retaining People

Towering technical competence begins with the system a company uses to hire, develop, and retain people. Many companies, unfortunately, do not have a system or a philosophy that supports towering technical competence with any consistency. They may have a few special engineers who excel at what they do simply by chance or some innate skill or drive. They may also acquire "the best and the brightest" through internships, at hiring fairs, or from universities. But all too often, these top-notch recruits are plopped into a gaping hole and left to develop as they will. With no process or program to guide them, they learn and develop haphazardly, often moving from one area to another too quickly or disjointedly to develop true expertise in anything at all. Further, since their supervisors were likely developed in the same way, they often do not have the experience and technical skill set to provide any meaningful technical mentoring to these new engineers.

A lean PD people system is a meritocracy and organized into a technical hierarchy created by developing and rewarding technical achievement. At Toyota, the career paths of all newly recruited engineers consist of practical hands-on work that develops deep technical competence. The engineers are evaluated regularly for six to eight years for demonstrated technical excellence, as well as process, and standards adherence. Rewards (advancement) are commensurate with progress and achievement. In many companies, management is the province of well-educated MBAs. At Toyota, the number of MBAs is conspicuously low. Upper management at Toyota consists of former engineers who revere technical excellence, seeing it as the true life-blood for product development. These managers have been developed through the same system and usually know a job better than the engineers reporting to them do. As a result, Toyota's "mentoring as leadership" principle works extremely well as it is perpetuated across generations of engineers.

How a company identifies its core competencies and values strongly influences its training and development practices. By viewing itself primarily as an *auto manufacturing* enterprise (a core competency), Toyota avoids many of the problems that other companies have in getting product engineers to design for manufacturing. Manufacturing engineering (which Toyota calls production engineering) is about as core as it gets. As a result of having this "identity" Toyota's engineers can do everything from designing tooling to developing complete production equipment and overseeing its construction, responsibilities that go considerably beyond what most manufacturing engineers do in most industries.

Hiring and training practices also create and influence the organizational culture (or value system) needed to sustain a lean PD system. This concept is best illustrated by example, and the example below revisits a company that should by now be quite familiar to the reader: NAC. This section provides an analysis of hiring/development practices at NAC and a contrasting look at hiring/development at Toyota. The authors suspect that many companies may see their own policies and cultural behavior reflected in NAC's policies.

Recruiting/Hiring Process at NAC

NAC has a decentralized hiring policy with each geographic/functional area responsible for its own recruiting/hiring. The numbers are based on a headquarters-authorized head count, which is incorporated into the functional area budgets. There is some central screening through human resources, which organizes recruiting visits to universities and then organizes visits of potential job candidates to NAC. Individual departments then get to decide on whom they want to hire. Once NAC places a new engineer in a department, he or she is "subjected" to the whims and capabilities of that particular department or manager—some may receive development and mentoring, others may not.

NAC engineers are encouraged to move quickly from department to department in order to "broaden their experience," which means any learning or mentoring tends to be rushed, chaotic, and unique to each individual. However, by staying in the same job for several years, an engineer risks being designated as deadwood: someone who is not going anywhere in the company. Those who need or want training may not get it and develop slowly (if at all). Others who possess a great deal of initiative "self train," develop quickly, and often become do-it-my-way "cowboys"— driving yet more variation into the PD system. NAC lacks a standardized people development process; there is no consistent career path, and there are no consistent criteria that determine evaluation and reward. What follows is an outline of NAC's process for recruiting and hiring for product engineering and manufacturing engineering as well as some common assumptions that many consumer-products companies make.

Recruiting/Hiring at NAC Product Engineering

To fill its product engineering ranks NAC traditionally recruits from the top U.S. universities, by participating in university-based recruiting initiatives. Typically, a large group of the best engineering students apply for positions at NAC. To be considered, applicants must have a bachelor's degree in mechanical, electrical, or chemical engineering. Many have master's degrees. They must pass a series of interviews in both the engineering and human resource departments. A successful applicant is usually hired by, and placed directly into the functional specialty for which he or she applied. Once on the job, new engineers are generally expected to become almost immediately productive and are turned loose on development projects with little in the way of structured mentoring. They learn in the "pressure cooker" environment of a new product program in a "trial by fire" methodology that produces a wide range of capabilities.

Hiring at NAC Manufacturing Engineering

To fill positions in manufacturing engineering, NAC follows similar practices to those outlined above. Although applicants are officially required to have a bachelor's degree in an engineering discipline, NAC will waive this requirement in lieu of experience. The manufacturing engineering group does participate in some recruiting activities although NAC usually hires people individually. The interview process takes place with both a manufacturing engineering manager and a human resources representative. The hiring process is not particularly rigorous, but NAC gives heavy preference to current NAC employees, transferring them into the department from other NAC activities (e.g. from assembly plants or stamping plants). Just as with product development, manufacturing engineering departments take responsibility for training new recruits and there is little standard protocol for this.

From the perspective of lean thinking, NAC's recruiting/hiring process is flawed and does not lend itself to creating a high-performance product development system. The company makes three assumptions that are detrimental to this process:

1. *Engineers are professionally trained in college.* By hiring top engineers from top engineering schools, NAC believes it is getting professionally-trained engineers and that the engineers have learned the foundations of professional engineering and can be productive immediately.

2. *Each department is capable of developing its own engineers.* Leaving the development of engineers to each functional department is perhaps the most dysfunctional assumption. In discussions with company managers involved in recruiting/hiring, the authors discovered that the prevailing attitude is "we are professionals so we can develop professional engineers effectively." Real-life experience strongly suggests otherwise. What's more, without a common beginning or company-grounded initiation process, engineers often develop more loyalty to their individual departments than the company as a whole.

3. *Manufacturing engineers do not require as rigorous technical and engineering training as product development engineers do.* This reflects the bias of NAC toward product engineering as higher status "real engineering." There seems to be a belief that manufacturing engineering is somehow less professional and is dirty work for the plant folks who do not need or have the applied math and science disciplines required for more challenging product development.

Training and Development at NAC

Employee development at NAC seems to revolve around breadth of experience. Almost as soon as they are hired, new engineers are told that broad experience is important to career success. The prevalent belief is that any one who stays in a job for more than a couple of years is not going to advance. Although there is certainly a significant opportunity for on-the-job training (OJT) at NAC, there is a lack of structure to the OJT process and mentoring and teaching are not rewarded behaviors for supervisors. Consequently there is little incentive for supervisors to develop new engineers. So while there are engineering supervisors who take great care in developing new talent they tend to be rare and their methods unique. Most formal training at NAC is online and often approached with a "check the box" mentality. For example, a series of classes is required for all NAC product engineering personnel. However, many of the engineers feel that the classroom training is essentially irrelevant and has little to do with how they do their jobs each day. Some of the training is considered valuable but does not change the way work gets done and engineers have expressed a desire for management to make some change in daily operations to reflect things that are learned in the classroom.

There is certainly a good deal of mentoring at NAC. However, it is typically not technical in nature, designed to grow proven technical skill sets. Much of the mentoring that we were told about was primarily about developing a career, focusing, for example, on how best to navigate the potentially hazardous political waters of the corporation. Though career-path development at NAC is often discussed, none of the people the authors interviewed could identify a particular career path for their respective functional specialty. Furthermore, they stated that people move around to different functions far too often to develop deep knowledge of any particular engineering discipline. Product engineers in particular are often moved into a new specialization before they have developed real competence in their current position.

Even in manufacturing engineering, where people do tend to stay in one place longer, there is still no prescribed career path for developing specific technical skills. One of the reasons for this is because much of the core manufacturing engineering takes place outside of NAC. Manufacturing engineering management must rely on what an engineer has brought to the job from his or her formal education, previous experience, or what they may have learned from suppliers. Without a valued and structured mentoring system and career path, NAC struggles to pass on standardized practices and tools to new engineers. This leads to a wide variation in the skills possessed by each engineer; some engineers are excellent, some of the best in the industry, while others are nearly incompetent, a situation that clearly contributes to high-task variation in the NAC process and detracts from NAC's ability to predict outcomes, plan, and standardize accurately. As noted in Chapter 5, high-task variation alone leads to queues and long lead times. An inability to plan and predict clearly exacerbates these problems.

Developing People at Toyota

At this point, the reader should clearly understand that creating a lean PD system entails not only adopting lean tools and a commitment to making the organization lean but also a change of philosophy about the way things get done. This is exemplified by the rigor with which Toyota selects and develops both product development and production engineers, using both OJT and mentoring in a fairly structured methodology. Right from the start, Toyota gives a high priority to the nurturing and development of engineering talent. In fact, people development at Toyota seems to be just as important as product development. Toyota managers are trained to be teachers and see every engineering project as an opportunity for developing its engineers. Developing people is fundamental to the manager's job. All managers view the performance of their teams as a direct reflection of their own ability. It is personal.

Hiring at Toyota

In Japan, a position with Toyota is highly coveted, and the company's hiring process is notoriously rigorous (note that it was not always like this and Toyota early on struggled to get good talent). Applicants from the best universities, such as Tokyo and Kyoto, far outstrip the available positions. Hiring is centralized, and engineers are typically hired in large groups that form a sort of freshman class each year. Classes are typically quite large and out of a group of 300 new hires, the distribution of educational level is approximately as follows: 2 Ph.D.s, 198 master's degrees, and 100 bachelor's degrees, all in an engineering discipline. At the time of hiring, new engineers are not certain where they will end up. Although Toyota will certainly utilize their respective specialties, the engineering organization in which they work is determined as they move through the initial segment of the career path.

Toyota's new engineering hires graduate at the top of their class in grades, but that is not the only criteria considered during the hiring process. Each engineering applicant goes through a series of intensive interviews designed to provide a comprehensive look at personal characteristics that determine whether a prospective hire will fit within the Toyota culture. Each candidate is thoroughly examined, and sources of information include professors who have a relationship with Toyota as well as current Toyota engineers. In fact, some recent graduates employed by Toyota as engineers are sent back to their universities to host recruiting parties in order to learn about applicants at many levels. Listed below are some key characteristics they look for in a new hire:

- Love of autos and technical work
- Technical capability
- Creative problem-solving ability (thinking outside the box)

- Teamwork (*nemawashi*, cooperate, share information)
- Ability to "grasp situation" quickly, thoroughly, and at a detailed level (what to look for, questions to ask, know what you need to know)
- Ability to communicate a situation succinctly
- Discipline to work consistently to a time schedule
- Motivation to work to targets
- Dedication to craft and company (e.g., willingness to work hours needed to get job done)

Training and Development at Toyota

The engineering career path at Toyota is commonly referred to as the "*Toyota T.*" It is an inverted (upside down) "T" with the top of the "T" symbolizing that an engineering career begins with a short time that is broad in scope and the base of the "T" symbolizing that it evolves over a long time within a specific technical discipline. The typical freshman class will undergo about one month of general training, including education on quality and orientation on the history and traditions of Toyota. Next new hires will spend time doing the manual work of building cars for about three to four months in a manufacturing plant. These engineers may then spend another two to three months at dealerships where they may be required to sell vehicles door to door. Toyota designed this training process to ensure that engineers understand the car business from Toyota's perspective as well as from the customer's perspective.

This common training regimen among each freshman class of engineers also gives new employees a sense of common beginnings, an important aspect of building a culture and instilling loyalty that lasts throughout a career. The message is clear: Each employee works for Toyota not for a specific function. During this orientation time, Toyota is constantly evaluating the new engineers to determine their best fit in the organization.

Training and Development at Toyota Body and Structures Engineering

When a new engineer comes to the body and structures engineering group, the department typically pairs him or her with a mentor (one of the senior engineers). Each new engineer is also assigned an improvement project, known as the "freshman project." This is usually a small but challenging technical project (such as reducing the required number of wiring clip holes in a part). The purpose of this exercise is to have the new engineer use basic engineering tools and seek out others to accomplish the task, a process that helps to teach about the Toyota way of engineering. For instance, one thing Toyota mentors emphasize is that an engineer should never just bring *the answer* to the boss. It is more important that new engineers consider the impact of several potential solutions and present these in decision matrix form or A3, thereby demonstrating that they have thoroughly considered the full situation. The freshman project continues the socialization process, further aligning the new engineer to the Toyota Way. Everything is new and the freshman project is challenging, a markedly different experience from university education and the general orientation of the first year at Toyota. For many new engineers, this experience can be quite emotional; it often makes such a strong impression that senior managers can still relate vivid details about their own freshman projects.

After new engineers complete their four- to nine-month freshman projects, Toyota assigns them to engineering specialties within the body and structures engineering organization. The engineers know that they must undergo a somewhat intensive two-tier OJT period (briefly described in Chapter 6). Engineers who successfully complete this OJT training earn a first-level engineering rank. The first tier, lasting about two years, is heads-down work at the computer-aided-design (CAD) terminal under the direction of a senior engineer (engineers must learn to do their own CAD work). After this initial first tier period, a body engineer can spend several more years (three to six) within this same technical specialty working on design of a few related body parts. Only then will management recognize this individual as an independently functioning lead engineer.

During the approximately eight-year development period, the manager continues mentoring and interviewing the engineer three to four times per year. The product of these interviews is not merely a manager's subjective opinion about job performance rating against peer performance. Instead, the mentor/manager uses Toyota's standardized skill set expectations to measure the engineer's technical progress and adherence to company processes and standard methodology. This information is supplemented by input from a variety of people the engineer has worked with. From these criteria, the manager outlines areas of improvement and develops an action plan that is measured in the next interview. Toyota also employs a *hoshin kanri* (policy deployment) process (discussed in Chapter 15), so that each engineer has specific objectives that he or she is being measured against. After being with the company 10 to 12 years, the engineer becomes eligible for promotion to a first-level management position.

Training and Development at Production Engineering

Like their counterparts in body engineering, new members of the production engineering organization come from the same freshman class and receive the same basic training in the first year. New production engineers are also assigned a mentor and a freshman project in the second year. In production engineering, this project might be something like decreasing the number of net surfaces or clamps required on a particular part fixture. In the stamping engineering group (part of product engineering), a new engineer's career path typically proceeds as follows:

- Four to six months in a freshman project
- Three to four years in die design
- Two to three years in processing and binder development
- Two to three years in the tryout and construction at Tool and Die

Unlike other companies (including NAC), Toyota does not have union constraints dictating who can do what, so Toyota engineers are free to participate fully in tryout and construction activities at tool and die. They also spend considerable time in the stamping plants as a matter of completing their tasks. Here too, new engineers are interviewed three times per year and evaluated against skills matrices.

After completing this development or apprenticeship period (about eight years), stamping engineers may be invited to join the simultaneous engineering organization, go to a stamping plant, or return to one of the stamping functional specialties such as die design. These assignments or deployments are planned with their managers, and the engineers can suggest their preferences. After several more years within their discipline of choice, they will be eligible for promotion to a first-tier management position.

For many companies considering lean PD, this significant investment in employees may seem unrealistic, particularly in Western companies with a high turnover rate of engineers. What these companies fail to appreciate is that this process of a lean PD system develops knowledgeable engineers while nurturing a vibrant lean culture. You can start this process by using two elements of lean:

1. *Standardization:* By having all departments use standard skill expectations to develop and measure new engineers, you are continually reinforcing and/or improving this skill set. Unlike many companies that have employees review standards once in a while in an online training class, you would actually apply standardization criteria daily in determining the competence of your workers. To appreciate this lean element's importance, remember that standardization helps reduce variability in the PD system and that *standardization leads to flexibility.*

2. *Learning Organization:* Being a leader in a lean organization also means being a teacher. Within a learning organization, a primary responsibility for managers is technical mentoring of engineers. This mentoring process also prepares a whole new generation of managers.

To augment the structured development process Toyota has a number of mechanisms designed for both organizational and individual learning and development. Toyota has repeatedly demonstrated a passion for learning and consistently builds opportunities to learn into the very foundation of their development process. We will have a detailed discussion of how Toyota builds learning and continuous improvement into their process to create a lean learning system in Chapter 11. There are also important implications for how these mechanisms support the development of towering technical competence in individual engineers.

Genchi Genbutsu Engineering

The phrase *genchi genbutsu* literally means the actual part, the actual place, but for Toyota, it implies *going to see the actual situation first hand to understand deeply the current reality.* It is one of the four core principles of Toyota's internal Toyota Way document and is manifested throughout the company (Liker, 2004). This hands-on approach to engineering is critical to a lean PD system and fundamental to developing new engineers. As Kiichiro Toyoda, founder of Toyota Motor Company, observes, "One can never trust an engineer who does not have to wash his hands before he eats dinner."

In this day of high-tech engineering, it is very tempting for engineers to divide their time equally between conference rooms and their cubicles. Moreover, in an environment that is increasingly more inclined toward overseas outsourcing and "virtual engineering," engineers may never get close to the product they are working on. But as Kelly Johnson, the famous head of Lockheed's legendary Skunk Works has commented, "An engineer should never be more than a stones throw away from the physical product" (Rich & Janos, 1994). This comment clearly reflects the spirit of *genchi genbutsu*. Examples of this philosophy in action include engineers spending preprogram time at dealerships, working on competitor teardowns, or personally fitting parts on prototypes. The CE's team also meets with customers, test drives vehicles, and evaluates its own and competitors' quality data. The main point of *genchi genbutsu* is that you can only develop quality products by having your engineers intellectually, physically, and emotionally connected to those products. The next few paragraphs describe some of the ways Toyota applies *genchi genbutsu* to product development.

Competitor Teardowns

In tearing down competitor's target products, Toyota identifies target vehicles and specific components as "best" in the new vehicle's class. The engineers then reduce these vehicles to individual parts and subject them to a competitive evaluation for quality, performance, and ease of manufacture. They mount and mark up the disassembled parts on a teardown board alongside Toyota's current model parts, displayed for all participants (including suppliers) to examine. In addition, they circulate the analysis reports among all participating functional groups for evaluation and comment. They now create these reports utilizing V-Comm technology, a virtual A3-type report that includes digital pictures, problem description, countermeasures and current product design geometry.

Prototype Builds

Body engineers practice *genchi genbutsu* during the prototype phase by participating in both virtual and physical prototype builds. They visit parts making sources and attend daily wrap-up meetings at the build site, often fitting and assembling parts that give them a true feel for what they have designed. Prototype phase is a time of intensive learning for the body engineer, and being at the source is invaluable. In addition to working with the SEs, body engineers work with the prototype specialists, Quality Assurance (QA) specialists, and production-assembly team leaders, who participate in the prototype builds. This phase, characterized by intense interaction, generates many engineering changes.

When engineering changes are deemed necessary during the prototype phase, they are often made on the spot where issues are identified. This is often at the build or parts making sources. Engineers mark up and sign sketches or drawings that serve as authorization for the changes to be incorporated. If changes cannot be made on the spot, body engineers must respond within 48 hours with new or additional data. This keeps the process moving forward while technicians are updating the actual product database. Throughout this process, Toyota achieves a performance advantage by using mutually supportive elements. This quick-change system requires experienced, knowledgeable engineers who understand the full implications of the decisions they make. They use tools such as checklists and decision matrices to reach rapid but quality decisions; moreover, the process allows engineers to work closely with the prototype build and parts fabricators and stay close to the product.

Daily Build Wrap-up Meetings

Another problem-solving and learning mechanism Toyota uses during its prototype phase is daily build wrap-up meetings. These are attended by the CE (or his staff), body engineers, prototype technicians, production team leaders, production engineers, and suppliers at the end of each day. Participants candidly discuss issues encountered during that day's assembly or incoming part inspection. The meetings are held right at the build site where participants can witness first hand the quality, cost, or productivity/ergonomic issue, and where they record issues/countermeasures and give new assignments on the spot. Here again, it is easy to see the spirit of *genchi genbutsu* personified by body engineers who work shoulder to shoulder with production team leaders and prototype technicians who crawl around the prototype vehicle, fitting panels and examining the results of their designs within the body system.

Your Lean PD System Must Develop People

Companies trying to adapt a lean PD system become particularly frustrated and concerned when they learn about the depth of knowledge and experience of Toyota's engineers and the number of years Toyota takes to recruit, train, and develop them. Because it requires a philosophical mind shift, this may be the most challenging aspect of the lean PD process for companies to replicate. And it can take much longer than most companies want to dedicate to implementing a lean system. But the philosophy behind this is the backbone of the lean PD system, supporting many of the LPDS principles in this book. Below are some of the reasons that Toyota's approach is culturally and historically unique, reasons that incidentally illustrate how one lean principle supports another.

- In the 1930s, Toyota's culture evolved out of a small company mentality where its engineers had to do almost everything hands-on from the ground up.
- Over time, this necessity created the cultural value of *genchi genbutsu*, the Toyota approach to solving problems.
- This behavior reinforced the Toyota belief in learning by doing.
- To excel at this, Toyota placed value on developing engineers with detail and deep expertise in focused areas.
- This value further supported the need for developing stable and standardized processes so Toyota engineers could work off detailed engineering checklists that reflect on-going learning.
- Because of the stability and homogeneity of past generations of Toyota engineers (homogeneity in terms of the strong culture of Toyota), a very similar set of expectations, values, and beliefs among leaders who were doing the mentoring evolved. The engineer truly learns the "Toyota Way" of engineering, regardless of who is doing the teaching.
- This reinforced Toyota's learning culture, which makes experimenting by trying things an everyday event and where mistakes are not punished.
- This "open attitude" encourages the spirit of innovation, putting engineers into challenging situations starting with the freshman project.

Of course, no company can replicate Toyota's approach or culture, but every company can adapt the lean thinking to begin to *Develop Towering Technical Competence in All Engineers*. The first step is to reinforce the premise that it is "people" that sustain the system and the more excellent the people, the more effective the system. And people need to be *developed*, not simply recruited or purchased. (Chapter 17 returns to this issue of building the culture and towering technical competence required for a lean PD system.) The next LPDS principle extends this behavior by emphasizing that companies should manage and nurture their suppliers in much the same way they do manufacturing and engineering resources.

LPDS Basics for Principle Seven

Develop towering technical competence in all engineers

People provide the energy and intelligence for any lean system, and product development is a particularly technical talent-driven enterprise. Consequently, LPDS requires that you focus significant time and energy in developing towering technical competence for all engineers. Start with a rigorous selection process, and then establish a technical mentoring system with regular evaluations that base assessment on demonstrated technical competence. The result will be significantly reduced task variation, powerful supplier management capability, and a level of professional trust that will enable lean product development speed. Toyota's culture honors technical capability and has created a technical meritocracy. This culture perpetuates technical excellence through mentoring, strategic assignments, and rigorous evaluations based on performance. Anyone who wants to create LPDS has to be prepared to make a significant investment in people selection and development.

CHAPTER 10

Fully Integrate Suppliers into the Product Development System

Achievement of business performance by the parent company through bullying suppliers is totally alien to the spirit of the Toyota Production System.

TAIICHI OHNO

DURING THE LATE 1990s, two U.S. auto companies came to the conclusion that their true core competencies were designing, assembling, and marketing cars. They believed they could style vehicles, purchase parts or "modules," bolt them together, and then sell and finance the car while parts suppliers competed for the components business. The functional idea behind this strategy was that outside suppliers would engineer most of the vehicle and manufacture virtually the entire car while the auto companies would do the final assembly. Even the engine was up for grabs. If someone else could make a quality engine and sell it cheaper, then more power to them. According to this model, megasuppliers would assume complete responsibility not only for manufacturing and basic engineering but also for the entire development of critical vehicle subsystems: seats, interiors, brakes, axles, and exterior sheet metal. The philosophical idea behind the strategy was to ask suppliers to take more responsibility, make greater technical contributions, and share risks "like they do in Japan." Naturally, this model had some attractive features for U.S. automakers to consider:

- Over time, they could shed a huge amount of fixed assets for manufacturing and make tremendous cuts in investment costs for engineering facilities, engineers, tooling, and equipment by pushing these onto suppliers.
- Suppliers were typically more cost effective than U.S. auto companies, due to more flexible work rules and lower labor rates than the union-dominated auto plants.
- Suppliers could develop greater specialized expertise, focusing only on those things they engineered and built.

- U.S. auto companies could use hardball price-negotiation tactics to force price reductions or give the business to another supplier.
- New web-enabled technologies for reverse online auctions would allow real-time pricing competitions between suppliers to achieve true world-class costs.

In short, this was a simple way for a company to reduce costs and at the same time become more agile and quicker—wonderful assets with which to enter the twenty-first century. When they compared themselves with Japanese companies, U.S. auto companies were making more than 50 percent of the components for any given vehicle while Japanese companies were outsourcing over 70 percent of their vehicle content. On surface, it made great sense to adopt the Toyota supplier-relationship model, but below the surface the plan was missing a few critical points.

A Part Is *Not* a Part, and A Supplier Is *Not* a Supplier

An automobile is a complex system. The way parts interact matters—even tires, mufflers, and glove compartments must be engineered to fit a particular vehicle. You must develop tools and dies, set up production lines, and manufacture the product to specification on time, with high quality every time. There is nothing trivial about any part or any of these steps.

When customers buy cars, they do not care who makes the engine, radio, seat, carpet, etc. They want and expect reliable quality and hold the automaker totally accountable for anything that is not up to their expectations. Toyota recognizes this and makes sure that every car part reflects Toyota quality. To achieve this, they make every supplier an extension of Toyota's PD process and lean logistics chain. Toyota delegates tasks to suppliers, but ultimately, it is Toyota that is fully responsible and fully accountable for all subsystems as well as the final product. Outsourcing does not absolve Toyota of responsibility or accountability.

General Motors, Ford, and DaimlerChrysler have, at different times, made genuine attempts to learn how to partner with suppliers by emulating the Toyota model. They have largely failed in these efforts because they have not grasped the concept of true partnering. When the market gets tough and there are serious earnings pressures, they have been less than forthright with suppliers, vacillating between public statements of commitment to partnership and trust and then unilaterally cutting supplier prices after signing contracts. They have even cut off suppliers in the middle of a contract, resourcing it to a lower bidder. Now and again, they have done so while using the designs of the first supplier. After a time, this behavior backfires and those who practice it earn a reputation as an untrustworthy business partner.

Surveys of U.S. auto suppliers consistently rate their supplier relationship with Toyota near the top. For example, a 2005 survey by John Henke's Planning Perspectives ranked Toyota number one on an index based on 17 categories from trust to perceived opportunity, followed by Honda and Nissan. Chrysler, Ford, and GM ranked fourth, fifth, and sixth respectively (Sherefkin and Cantwell, 2003). Toyota had its best showing scoring 415 out of 500 (32 percent better than 2002) while GM was at an all time low with a score of 114 out of 500 (29 percent worse than 2002). A survey conducted by J.D. Power & Associates found that Nissan, Toyota, and BMW were the best North American automakers in promoting supplier innovation (*Automotive News*, Feb. 24, 2003). Honda and Mercedes also finished above average in fostering innovation whereas the Chrysler group, Ford, and General Motors all were rated below average. There are obvious reasons why suppliers rank Toyota so highly:

- Works with new or struggling suppliers to get up to speed.
- Makes commitments to suppliers early in the product development process and makes good on promises.
- Constructs contracts that are simple and for the life of the vehicle model.
- Is the best at balancing a focus on cost with a focus on quality compared to the other automakers.
- Honors the contracts—does not renege on them to save cost.
- Treats suppliers respectfully and respects the integrity of intellectual property.
- Sets aggressive price reduction targets, but works with suppliers to achieve the targets.

This does not mean "easy street" for suppliers. The suppliers the authors interviewed consistently rate Toyota as *their most demanding customer*—demanding in timelineness, innovation, quality, and cost reduction. For example, when Toyota discovered the prices it was paying suppliers for many key parts were above the lowest prices competitors paid globally, it issued a new target for all key suppliers as part of a program designated CCC21. The new program asked suppliers to reduce prices by 30 percent for the next model line; typically this would cover a period of about three years. Instead of rebelling, suppliers agreed to the new terms. Why? Because Toyota agreed to work with them on cost reduction, even offering to help suppliers change the product design through value engineering. Because it responds to supplier concerns with integrity and capability, Toyota has established a great level of professional trust with its suppliers, closely paralleling the trust that Toyota engineers have established with each other.

As the 13 principles of our LPDS model show, Toyota has developed a sophisticated system that encompasses people, processes, and tools and technology and extends to all partners in the lean enterprise. As Toyota partners, suppliers must follow the same or equivalent processes for design and manufacture. Moreover, once accepted into the Toyota family, they are taught to be effective partners. Cost reduction is one outcome of the selection/teaching process. Toyota is a master at cost reduction internally and expects suppliers to master this discipline as well. In a lean PD system, you cannot "bully suppliers" and squeeze the life out of them for a price reduction. The bottom line, though always important, does not drive supplier relations. Developing excellent processes and quality products is the goal, and to achieve this, companies need to extend the learning enterprise and truly partner with their suppliers.

The Power of the *Keiretsu*

In the 1980s, the authors visited Japan to learn about the highly publicized *keiretsu* (set of interlocking corporations). In the *keiretsu* model, a broad set of different types of companies cooperate in business and hold equity in each other. Automakers hold equity in a close-knit group of affiliated suppliers that are essentially part of the company. Under this arrangement, suppliers were told which companies they could do business with; because the affiliated automakers controlled this, they were able to trust the suppliers with sensitive information, the kind of information generally kept secret and guarded even within individual companies. What the authors discovered in Japan, however, was that the supply chain in this model looked more like a hierarchy of interlocking corporations rather than a chain. A more accurate description of *keiretsu* then is that it is a hierarchical organization that does a great deal of parts outsourcing to small numbers of closely-knit, large volume suppliers who are given long-term contracts and are at different levels of the hierarchy (Kamath and Liker, 1994). There is competition among suppliers, but typically two or three suppliers make a given type of part and have 100 percent of the business for a given vehicle program. Toyota selects these suppliers early in the product development program, guarantees the business, and incorporates them as part of the extended product development team.

Are All Suppliers Created Equal?

Toyota uses a tier structure for its suppliers, similar to a hierarchical organizational chart. It deals primarily with the first tier, the largest suppliers who provide complete subsystems directly to Toyota. That tier manages the second tier of component suppliers that ship parts to the first-tier plants[1] and so on. This makes the management task more efficient because Toyota does not have to communicate directly with thousands of suppliers. To further manage the *keirutsu*, Japanese suppliers have four major roles for Toyota's different vehicle programs, each of which are discussed below (see Figure 10-1) (Kamath and Liker, 1994).[2]

1. *Partner*. This is the highest level, which includes companies like Denso, Araco, and Aisin. These companies have grown to be comparable in size to Toyota and are technically autonomous. They can design their own subsystems and components and have complete prototype and test capabilities. They are involved in the earliest concept stages at Toyota, and they often develop sketches before Toyota has created a contract or even developed formal specifications for the subsystem. Partners have a large number of "guest engineers" housed in Toyota's design offices near Toyota engineers for 2- to 3-year stays. These engineers help augment Toyota's engineering workforce without increasing Toyota's payroll; they collaborate on the design and solve design problems. While the detailed design goes on back at the supplier's engineering organization, the guest engineers serve as key liaisons. By cycling engineers through Toyota as guest engineers, suppliers are also training their workforce in Toyota's product development system.

---

1. There are exceptions to this. For example, for some basic raw materials like steel, Toyota uses its purchasing leverage to deal directly with the suppliers and get a better price based on its large volume and deals with these suppliers directly though they are not first tier.
2. The details of these roles were first put together by Durward Sobek (Montana State University) based on a research trip to Japan.

	Contractual	Consultative	Mature	Partner
Design responsibility	Customer	Joint design	Supplier	Supplier
Product complexity	Simple parts	Simple assembly	Complex assembly	Complete subsystem
Customer specifications provided	Complete design or supplier catalogue	Detailed specifications	Critical specifications	Concept
Supplier influence on specifications	None	Present capabilities	Negotiate	Collaborate
Timing of supplier involvement	Prototype	Post-concept	Concept	Preconcept
Component testing responsibility	Customer	Supplier input	Joint	Supplier
Supplier development capabilities	Little	Significant	Strong	Self-contained

Figure 10-1. Maturity of Suppliers and Roles in Product Development

Source: R. Kamath and J. Liker, "A Second Look at Japanese Product Development," *Harvard Business Review*, Nov.–Dec., 1994: 154–173

There are only a few suppliers that fit the image of a full partner, and even these suppliers must continually prove themselves.

2. *Mature.* A majority of first-tier suppliers are just a half step below the partner level. They have matured to the point that they have very strong engineering and manufacturing capabilities, but they are a bit less autonomous and depend more on direction from Toyota than partners. Their products are not quite as complex, and they rely on specifications from Toyota. It is here that Toyota's set-based approach to setting specifications differs from the approach used in U.S. auto companies. Whereas U.S. companies set detailed and rigid engineering specifications for suppliers, Toyota provides less restrictive specifications and views them as targets (Ward et al, 1995). When delivering these somewhat vague specifications to suppliers, Toyota often uses the word *gurai*, the Japanese equivalent of "about." One of the difficulties Toyota has had working on engineering tasks with U.S. suppliers is that these suppliers need explicit instructions, including detailed specifications and tolerances before they can act. Worse yet, if Toyota does not specifically ask the supplier for something, it is not given. This has been a frustrating experience, particularly for Toyota's Japanese engineers who are used to working with Japanese suppliers, who often anticipate their needs. Mature suppliers do not wait to be asked. They take initiative and suggest. *Toyota wants the suppliers to think for themselves, challenge the requirements, and provide value-added ideas to the process.*

3. *Consultative.* These suppliers make commodities like batteries and tires, but Toyota also taps into their expertise, getting ideas for products still being decided upon. Consultative suppliers influence the specifications by recommending their own innovations, for example, a new tire with new characteristics. These products are generally not technically complex, and there is less intense engineering collaboration with this supplier group except around the prototype testing and launch stages.

4. *Contractual.* Toyota purchases nuts and bolts, brackets, and spark plugs—parts that do not require a major partnership. In many cases, Toyota simply specifies what it wants, picking from a catalogue or engineers a special component, if needed, and then selects a supplier. First-tier suppliers have a host of these commodity suppliers, but even with these contractual suppliers,

Toyota closely monitors quality, cost, and delivery. Toyota seeks out contractual suppliers that deliver just-in-time shipments, parts packaged in the right quantity and right container, and only the best quality. These suppliers are also expected to work diligently on cost reduction.

Toyota teaches the Toyota Production System to suppliers at all levels, chooses its suppliers carefully, and is cautious about which suppliers it lets into its extended family and to what degree it becomes intertwined with its partners.

Selecting and Developing Toyota Suppliers to Partner: U.S. Supplier Tire Example

The core *keiretsu* partners of Toyota in Japan have been part of the company's extended enterprise for decades, so it is difficult for any new suppliers to break into the network unless they have patented technology that Toyota wants. Outside of Japan, in the early stages of launching production, however, it is a different story: Toyota works to grow its local supply networks.

One example of this is the tires developed by suppliers working with the Toyota Technical Center in Ann Arbor, Michigan, for North American cars. A tire seems to be a pretty simple part of a vehicle that can be easily sourced just on the basis of price listings in a catalogue. Not according to Toyota. The tire is an intricate part of the chassis and a key to controlling noise, the feel of the ride, handling, safety, and even fuel economy. When the chief engineer specifies his or her performance expectations for a vehicle, e.g., handling and ride comfort, chassis engineers tweak the suspension and the tires to meet these objectives. The stopping distance is a function of the tires and the brakes. If in optimizing ride comfort and handling, the tires cannot withstand frequent stopping, the chassis engineer must make up for it in the brake system. Toyota works through a detailed process to specify, solicit, and qualify tires that are suitable for each vehicle's unique properties and tire specifications. What this means is that a tire supplier has to take into consideration many technical issues, a lengthy process that requires a strong supplier relationship.

The relationship between suppliers and Toyota's expectations of the supplier is very complex, even when it involves only one specific car component, such as a tire. Because Toyota deals with only two or three suppliers in a region, the company does not frequently bring on a new supplier unless there is an unfilled need or an opening. A new supplier gets a very small amount of business and is tested rigorously before getting any more business. One aspect of this testing involves investing in the development process as described below.

1. *Tire R&D.* Typically, tire development starts with the supplier who is continually doing R&D to develop new tires. The supplier must consider appearance, competition, performance requirements of the market, manufacturing capabilities, cost, government regulations, and where it wants to market (e.g., to automakers or just to aftermarket). The supplier develops a basic tire profile by looking at different tire compounds—down to the trees the rubber will come from (e.g., rubber trees in Vietnam are different from those in India) and considering the tread pattern (affects durability, wet and dry traction, hydroplaning, noise, cornering). Variables in tire construction (e.g., number of belts and of angles and materials of belts) are also thoroughly examined. The supplier then performs computer simulations and physical tests to assess other concerns (e.g., strain on tire, how much movement tire elements get under pressure).

2. *Setting requirement for the tire.* A Toyota Motor Company tire committee sets Toyota-wide standards for size, rolling resistance, etc., and the chief engineer sets the tire requirements for the development of a particular vehicle. These include objective requirements (e.g., stopping distance) and subjective requirements (e.g., the turning ability and handling). Because Toyota is committed to becoming a "green company," the CE also considers fuel efficiency. Often the specifications are relative to the current model or a competitive vehicle with better cornering and handling or better road feel for the driver. Toyota engineers benchmark a variety of tires to see how they meet these specifications, including tires used on competitive vehicles. They will also consider tires presented to them by their suppliers, sometimes on the basis of new technology. Toyota tests the tires in existing vehicles, looks at customer feedback from the dealer network, and considers data from J.D. Powers, Consumer Reports, Car and Driver, and other sources. The company also checks on current and pending government regulations.

3. *Request for Design and Development Proposal (RDDP)*. Toyota runs the specifications into the RDDP document, which is a formal request given to a set of selected suppliers to codevelop and quote on the tires. The document typically has at least four pages of detailed specifications, presented in the order of importance and based on a specific vehicle (e.g., 1999 Camry, AC, ABS, no sunroof, and four doors). The quantitative targets for such things as mass, speed rating, inflation pressure, and rolling resistance are numerical. They also provide subjective requirements in relation to a control tire Toyota has tested. For each subjective dimension, a chart shows where the control tire is (green dot), what the target is (red dot), and if it is different from the control tire (for example, better grip, less noise when it hits a bump on the road, cornering). One column in the chart indicates who is responsible for testing conformance. Toyota performs tests on an actual prototype, while the tire maker performs tests on specialized equipment. There is also a comments column in the chart. Toyota also gives the supplier a target price, and there is very little room for negotiation.

4. *Awarding the business to a supplier*. Engineering works with purchasing to select a suitable supplier or suppliers, always including one more supplier than Toyota is planning to use. Assuming Toyota needs two suppliers, three suppliers are invited to submit a quote. Typically, these are suppliers that already have a proven track record with Toyota. The selected suppliers *are expected to codevelop the tires with Toyota, using their own funding.*

5. *Reviewing and testing the prototype tires*. Toyota expects the supplier to develop two tire alternatives for the first prototype submission (1S). Sometimes this is a brand new tire, but most often, it is a derivative of an existing tire with one or more modifications. (The supplier, for example, may change nothing more than the tread pattern.) Toyota will answer questions, test tires at the Toyota test track while under development, and work with the supplier throughout the process. At 1S, engineers thoroughly evaluate the two tires by reviewing data from the supplier and on the test track. Toyota will then select one of the tires, or ask for some combination of the two tire alternatives, or ask for additional improvements. If there are difficulties meeting the CE's targets, the chassis engineers consult with the CE. The CE determines if

the tire is "close enough" or whether the engineers need to continue working toward a solution.

The supplier then provides a single tire at the second prototype stage (2S) for evaluation. The tire development proceeds in tandem with the development of the overall vehicle prototypes, but this does not necessarily occur in lockstep. However, the tires at prototype stage 2S must meet vehicle requirements and be available on time for the final confirmation build of the vehicle prior to launch. For initial development, Toyota works with the prior year's model; final development is done on a prototype of the new vehicle.

Toyota conducts tests on its North American vehicles at the Arizona proving grounds near Phoenix, which has test tracks accurately simulating different road conditions, including exact replicas of segments of I-94 in Detroit (based on measurements from a typical day), and a highway outside Los Angeles. Dry tests, wet tests, and tests on ice are all conducted. The testers take the vehicle to the limit, for example, turning on severe curves. Toyota measures actual stopping distance, and if the tire does not meet the requirements, the supplier is asked to make modifications.

6. *Choosing the final supplier(s).* After the 2S build, Toyota drops the supplier that does not meet the performance requirements. If all three meet the requirements, Toyota purchasing will consider other factors when deciding which supplier to drop such as who is the incumbent for this car model. Once Toyota selects the supplier(s) it prefers, it continues to monitor quality and delivery performance through the initial months of launch to make sure both continue to meet Toyota standards.

Every new supplier must go through all these steps and make required investments in the development process with no guaranteed return on investment. New suppliers, when first chosen by Toyota, would generally not get tire business for a major new model but might supply only the spare tire for a lower volume vehicle. Typically, it takes a supplier several programs and years of engineering effort to get enough business from Toyota to start collecting on the initial investment. Nonetheless, the competition to become a Toyota supplier shows that the relationship with Toyota is highly prized.

Partnering with Suppliers: Who Gets What?

The tire example above shows Toyota's rigorous process for developing and selecting components and suppliers and is indicative of a lean company's expectations of its suppliers. But how does this partner paradigm compare with traditional models of automaker-supplier relationships? There are, of course, some similarities, but the differences (presented below) underscore why a lean approach within the relationship supports design, manufacturing, and, ultimately, profitability.

Suppliers Work Closely with a Company: Mutually Beneficial Long-term Relationships

Traditionally, U.S. suppliers have not worked closely with U.S. auto companies in the development stage. For suppliers working with lean companies, this is imperative. They must not only learn how to deliver on requirements during this stage but must also pay thorough attention to various detailed requirements and offer alternative solutions.

Gotsu-gotsu is a Japanese term that refers to how a tire responds to a low frequency, high impact bump. The driver or passenger feels this impact in the lower back. *Buru-buru* is the tire's reaction to a low frequency, lower impact that the driver or passenger feels in the belly. When Toyota engineers say, "There is too much *gotsu-gotsu*," the suppliers generally understand what that means and how to respond. In Japan, this is taken for granted. In the United States, suppliers need to learn these and similar terms to participate effectively in Toyota's product development. Toyota uses a deliberate step-by-step approach to teach terminology and requirements, developing suppliers slowly and steadily.

This gradual development approach can be frustrating for suppliers, especially since Toyota requires them to invest in development and set up manufacturing lines. When Toyota was developing one vehicle in North America, for example, a new supplier developed reinforcement beams over a two-year period. The supplier did a good job developing and was awarded the contract for the vehicle in question but lost its bid for other contracts that followed. Thus, the supplier's development cost was a hefty investment leading to a relatively small yield. Eventually, however, the supplier's standards improved to a level that matched Toyota standards; more work was awarded and the supplier gained the stability of a long-term partnership with Toyota. In another example, Toyota brought on a new supplier to develop a new type of energy management system. The material the supplier developed for regular injection molding worked better than expanded foam that was used previously. Toyota, however, spent four years analyzing and evaluating the new foam material and energy management system before committing to a small contract. Because this particular material had never been tried in a production vehicle, Toyota waited until the supplier secured a contract with another customer to prove it was viable in actual production before finally committing to a contract.

New U.S. suppliers seldom meet Toyota's expectations at first, and it can be challenging for both parties to work together. For example, for body fit and finish, automakers develop tolerances for individual stampings based on stacking analysis. This requires data from the supplier that shows what tolerances it can hold. One young Toyota engineer received tolerance data for stamped parts from a supplier, but the means and variances were the same for many different dimensions. Toyota had requested coordinate measurement data for 1,000 parts. The engineer knew that what he was looking at could not be right and was convinced that the supplier must have fudged the data. He went to the supplier's plant and helped revamp the data collection system. The accurate data showed that the plant had not met the tolerance criteria, so the supplier had to do a thorough 5-why analysis to figure out the source of the variance and correct it. The Toyota design engineer developed the supplier by teaching the proper way to collect accurate data, analyze it, and develop corrective actions to improve quality in manufacturing. In the traditional supplier-company relationship, this seldom happens. Auto companies in this model also require data, but they seldom analyze it as thoroughly, and when they do find inconsistencies, they usually punish suppliers rather than teach them. In a lean PD system, you teach suppliers that are willing to learn and improve, and this enables a valuable, continuous partnership.

Price Is Not Everything

In a typical supplier relationship, once certain quality standards are met, price becomes the company's overriding consideration in choosing a supplier. In lean, the supplier needs to meet the performance requirements and the price targets but must also show an ability to partner with the company day by day as details are smoothed out and concerns are addressed. Toyota as a rule does not source on the basis of price alone. In the tire example above, GM, Ford, and DaimlerChrysler all announced aggressive targets for sourcing from China for American models to capitalize on the low labor rates. Bringing in China suppliers often means setting a spec taken from the data of current suppliers, testing some tires for conformance to the spec, and perhaps examining source plant facilities. If the company is satisfied, and if the price is lower, the supplier gets the business. Darrel Sterzinger, general manager of engineering design, chassis, at the Toyota Technical Center comments on this:

> That would be unthinkable at Toyota. I could not sleep at night if I did that. If we did bring on a supplier from China for some reason, they would have to go through the same codevelopment process as other suppliers. They would have to compete in this process. And we would start off by giving them a small job, like the spare tire on a low volume vehicle. We would monitor their performance. If they performed well on product development, quality, and delivery we might give them the spare tire on a higher volume vehicle. And then, if they performed well over several years, they might get the tires for a larger volume program.

Suppliers have burned Toyota with promises of low-cost alternatives. In one case, a company's prices were so competitive that purchasing awarded it business for a rear tail lamp before engineering thought this supplier was ready. It seems the company wanted the business badly and was relocating the business to Mexico to get the benefit of lower labor rates. Once the company launched manufacturing, the scrap rates were off the charts by Toyota standards. Toyota engineers tried to develop the supplier but were unable to make enough headway. It became clear that the supplier simply could not meet Toyota standards and requirements. The supplier was taken off the project, and the relationship was dissolved. The lesson was that *it is more expensive in the long run to pick the cheapest supplier if this supplier is not ready to meet your requirements.*

Losing a Bid

In traditional relationships, a number of suppliers bid on a contract and someone wins. Not getting the contract may be painful for a supplier, but because little was invested, little has been lost. In the lean approach, a small group of suppliers competes throughout the development process.

Although each supplier makes a serious investment in R&D, some win and some do not. But suppliers with an understanding of lean see the R&D investment as an overall investment in a relationship: Losing a particular program does not mean losing the investment. In fact, suppliers that work with Toyota may lose several bids before they win lucrative contracts. Most suppliers begin a partnership with Toyota that is paced incrementally and is characterized by losing large contracts and winning small contracts until Toyota decides that the balance can be reversed.

Relationship Development

As noted above, suppliers who do not satisfy a traditional enterprise on a particular job are seldom given the opportunity to work for that enterprise again. In contrast, a lean enterprise views supplier relationships in light of development potential. This is particularly true at Toyota.

Toyota puts all new suppliers through a thorough, step-by-step regimen. For example, when Toyota first started working with General Tire, the relationship began on a very small scale. Toyota was first tried out as a supplier of spare tires on the *Camry* station wagon—a low volume vehicle. The company then adapted this as a spare for the *Sienna* minivan. Toyota constantly monitored performance, teaching the supplier the Toyota approach to product development and manufacturing. At times the company failed to meet Toyota's standards, but Toyota continued to teach as long as the supplier was eager to learn. General Tire's work was eventually expanded to supplying all four tires for the *Avalon* and *Solara*—still relatively low volume vehicles. Toyota then considered using the company as a supplier for high volume trucks and SUVs. Gradually General Tire became a valued and trusted Toyota supplier, *but it took ten years*.

The Guest Engineer System

Toyota always has a crew of hundreds of engineers from suppliers, residing full time in its product development office called guest or resident engineers. While they have separate areas, they interact daily with Toyota engineers. This has an obvious benefit—free engineering resources for Toyota. But that is not the purpose. The purpose is integration.

When Toyota invites a supplier to send guest engineers, it is a significant commitment to long-term coprosperity. The supplier knows they have earned a long-term place in the Toyota enterprise. These are highly coveted positions. The supplier will become intimately familiar with Toyota's product development practices and get advanced information on new model programs.

The supplier can also realistically assume that engineers with a Toyota experience will become better engineers. It is like sending employees on the company payroll to a top engineering school—perhaps better. They will learn the Toyota Way of engineering. They can bring what they learn back to the supplier, enhancing the product development processes of the supplier. This of course assumes that the customer has something to teach, which in this case, with Toyota as the customer, is a good assumption.

Guest engineers rotate through this position usually in two- to three-year stays. Large suppliers like Denso have many engineers at a given point of time, so over the years, many of Denso's guest engineers will have had this deep Toyota experience. When we speak of how interchangeable Toyota engineers are with those of key suppliers enabling flexible capacity, the guest engineer system is a main reason for this.

The Supplier Stable

A traditional enterprise often relies on a selective group of favorite suppliers, but because the primary purpose of the relationship is to acquire needed supplies at a good price, such an enterprise does not hesitate to widen its supplier base. If the existing suppliers are outbid or cannot meet particular requirements, the company looks elsewhere without a backward glance. In many cases, the relationship with the original supplier is permanently terminated.

A lean model, on the other hand, generally works with a small stable of suppliers within the context of a broader plan. The goal is to keep these suppliers busy and competitive and to motivate them to do their best, sometimes with an eye to what they may contribute to other programs in the future.

With respect to this broad and long-range philosophy, Toyota's purchasing department plans and manages how much business each supplier will get on average over time. If a supplier is performing well but is not selected for one particular program, that supplier is likely to be selected for another program in the near future. If a supplier has performed poorly on quality, product development, delivery, or meeting price targets, that supplier may lose some share of the purchasing department's proposed division of contracts to be awarded but has the opportunity to regain the lost share. Because Toyota is willing to work with suppliers to help them improve, the suppliers are likely to work hard to meet Toyota's standards and requirements. Ultimately, they win contracts for other programs.

The Crux of Outsourcing Strategy

As mentioned earlier, Toyota successfully outsources over 70 percent of its vehicle content and this seems to have encouraged other auto companies to do likewise. But in trying to emulate Toyota's success, most of these companies do not fully appreciate or understand the strategy that makes Toyota succeed: Although Toyota outsources parts and even engineering, *it does not outsource competence or relinquish control.*

Like any company, Toyota sees many advantages in outsourcing, including the flexibility of putting expert supplier engineers on PD program teams to work on specific components as needed. However, Toyota is careful when deciding what to outsource and what to retain in-house. Even with outsourced components, Toyota is not willing to give up internal competency, even though it may be cheaper or more convenient to do so.

Mastering Core Technology

Maintaining a core competency is a current management buzz phrase among auto companies, but Toyota has its own unique interpretation of this concept. Rather than seeing itself as just designing, assembling, and marketing cars, Toyota defines its competency around the technology of selling, designing, engineering, and especially manufacturing of transportation vehicles. Moreover, when Toyota outsources, it does not relinquish control; it wants to learn and excel in new technology along with suppliers. While it hands off certain responsibilities, Toyota *never transfers to its suppliers all of its core knowledge or full responsibility for any core area.*

When new technology is core to a vehicle, Toyota distinguishes itself from the rest of the pack by diligently working to master the technology and becoming the best in the world at using it. As a lean organization, Toyota is fully aware that mastering any core technology internally helps to 1) manage suppliers effectively (e.g., understand true costs), and 2) continue to learn as an organization to stay at the forefront of the technology.

Chapter 7 described in some detail the *Prius* hybrid vehicle development, which depended on fundamental technology that Toyota had never before engineered into a mass-produced vehicle. In particular, there were three key technologies never before developed in-house: 1) the hybrid's electric motor, 2) the powerful battery, and 3) computer controls (IGBTs) that switch the battery's DC current to a different form.[3] Because of the short timeline to launch the *Prius*, a natural response would have been to partner with outside companies with expertise in these three fundamental technologies. Instead, even though President Okuda continued to move up the release date, senior management and the *Prius* CE, Uchiyamada, insisted on developing as much of the hybrid car in-house as possible.

Developing new capability: the hybrid electric motor and computer controls

Toyota's senior managing directors saw the hybrid electric motor and, specifically, developing capability in applying the Insulated Gate Bipolar Transistor (IGBT) to autos as an opportunity to corner the market on components for future hybrid vehicles. IGBT are switches that convert DC current to three-phase current and are key to the ability to move back and forth between the electric motor and the traditional gas engine. As a result, Toyota developed the IGBT in-house and, in the process, started a new semiconductor manufacturing business. The development process was demanding, conducted by engineers with no previous experience in this technology, and Toyota invested 5 billion yen in the new plant. The return on this investment is still paying benefits. Today, Toyota outsources many of these parts originally developed in-house, but Toyota has a degree of control over the supply of these parts and controls cost. Moreover, Toyota now knows how to produce small runs economically and to reduce costs for these small batch parts over time—a general lesson it can apply to niche vehicles.

Outsourcing the battery while maintaining capability

Though Toyota sorely wanted to develop new technology around the battery, there was a constant problem with the part driving the electronic motor portion of the hybrid. In the end, Toyota had to outsource this component. But rather than simply handing off the responsibility to a supplier, Toyota established a joint venture company with Matsuhita— Panasonic EV Energy (Itazaki, 1991). Having worked with Matsushita

---

3. The story of how Toyota developed these three components are discussed in detail in the book, *The Toyota Way* (Liker, 2004).

before, Toyota was somewhat confident about the relationship, but this confidence was tempered with a concern about differences in corporate culture. Of specific concern was whether Matsushita had the proper discipline to handle the quality control issues necessary to develop a completely new type of battery. Fujii, Toyota's engineer responsible for the battery, eventually realized that there was a way to meld the style of the two companies to address this issue. Thus, what began as a third-tier consultative relationship eventually evolved into a second-tier mature relationship as Toyota engineers worked hand in hand with their joint venture partner.

Changing Policy to Maintain Internal Capability

Even when it chooses to outsource a key component, Toyota prefers to maintain internal capability and competency. Consider, for example, its long relationship with Denso, which was a division of Toyota until becoming an independent enterprise in 1949. Denso is one of the largest global parts suppliers in the world, and through the *keiretsu*, Toyota now has controlling interest with about 20 percent of the company. Over the years, Denso has been Toyota's electrical and electronic parts supplier of choice despite its close relationships with Toyota's competitors. As a rule, Toyota wants at least two suppliers for every component, but Denso, operating like a division of Toyota, has often served as the company's sole electronics' supplier. This arrangement, though usually smooth, sometimes led to conflict, but neither company seemed inclined to change it.

Then, in 1988, seemingly out of the blue, Toyota opened an electronics plant in Hirose and began aggressively recruiting electrical engineers. The compelling reason for this change in policy was that Toyota had recognized that electronics technology was becoming a prominent part of vehicles. (Today as much as 30 percent of the vehicle content is electronics related.) Even after years of outsourcing to Denso, Toyota was capable of changing course and initiating a learning-by-doing program to "infuse its entire organization with the skills and values essential to making electronics a genuine core competence" (Ahmadjian and Lincoln, 2001). At this time, about 30 percent of Toyota recruits are electrical engineers.

Using Keiretsu to maintain internal capability

As discussed earlier, Toyota maintains internal capability through its *keiretsu* by having first-tier partners and mature suppliers that take major responsibility for subsystem engineering and testing. Toyota is still dependent on its suppliers because they have dedicated assets, like tooling designed for a certain car model and in-depth product engineering knowledge, which is difficult to duplicate. For this reason, Toyota maintains control through direct ownership of a portion of each of its suppliers and on interlocking boards of directors. In its move toward doing more business with U.S. suppliers, Toyota (like the big-3) has insisted that first-tier suppliers set up divisions dedicated to serving only Toyota. Furthermore, to sustain the integrity of proprietorship, these suppliers are required to build and maintain firewalls with any of Toyota's competitors that they may also be serving.

Using Keiretsu megasuppliers to build modules

One of the trends in the auto industry (as well as in many consumer-driven businesses) is modularity. Like other products, cars can be divided into a set of interdependent modules (interior cockpit, corner module, etc.), each of which can be sourced to an outside supplier to engineer, then built, and shipped in sequence to the assembly line. The automaker bolts together the modules and the work is done. Toyota, preferring to keep more control internally, was especially conservative about joining this trend. Instead, Toyota created 13 or 14 *keiretsu* megasuppliers and, by 2003, it had developed its own unique strategy for modularity, one that gave it a degree of control and internal competency. These new Japanese megasuppliers have very broad technical and program management competencies and take on responsibility for design, development, production, and assembly of modules. It is important to note here that the megasuppliers remain part of the *keiretsu* network and that Toyota as the OEM maintains redundant competencies to oversee their operations and development. In addition, by establishing new joint ventures with these suppliers, Toyota takes a substantial portion of the generated equity.

On surface, there is an inherent contradiction in supplier relationships in the lean PD approach. On the one hand, when you do business with any outside company, you need to treat them with the respect accorded a partner. Moreover, when suppliers become part of the inner workings of your company, they then become part of the extended network that make up your enterprise. Trust and openly sharing information are critical to success. On the other hand, it is important to define your core competency properly and accurately, and then aggressively find ways to maintain it. You also need to identify and monitor internal capability, a process that sometimes prompts starting a new business or changing long-held policies—both a huge investment of resources. On some levels, Toyota may seem unduly conservative and protective of its internal control. But it is because of this that Toyota has managed to maintain a balance between two apparently conflicting ideas by applying one invaluable principle: *hold outside partners to the same high standards as inside engineers and, until you build trust, be reluctant to relinquish control.* The level of trust will vary for each company. For Toyota, even when trust is established, there is still clear difference between being inside and outside of Toyota, and Toyota always reserves the right to keep core technical competence to engineer and build key components in-house.

Treating Suppliers Respectfully and Reasonably

When a U.S. supplier was asked how Toyota ranked as a partnering company, he answered with no hesitation that Toyota was his best customer, following up with an example that explained his answer. He said that when his other customers place an order in advance and decide they don't need as much as they ordered, they ask him to eat the cost. When Toyota adjusts its order lower, it purchases the total number of parts originally requested. When Toyota makes a commitment, it honors that commitment. In other words, it plays fair.

The supplier had another interesting observation. When he deals with Chrysler on a new launch, the company pulls out a quality procedures manual the size of Webster's dictionary. To make things worse, the manual is constantly changing and keeping up with the changes consumes a lot of time and human resources. This is typical and underscores why many suppliers think that their big-3 American customers use bureaucracy, especially "quality" bureaucracy, as a tool to beat them up. Time after time, confused or incomplete information from the customer turns into higher costs and lower profits for the supplier.

Both U.S. auto companies and Toyota use bureaucracy—extensive standards, auditing procedures, rules—in their day-to-day business with suppliers. However, many suppliers view U.S. auto companies as highly coercive, inconsistent, and incompetent. On the other hand, they see Toyota as an enabling partner—demanding, but willing to work things out together. Toyota engineers have the skills and experience to understand critical issues and provide simple clear direction. They do not have to hide behind layers of unnecessary administrative requirements or bureaucratic jargon. One U.S. automotive interior supplier described working with Toyota in this way:

When it comes to fixing problems, Toyota does not come in and run detailed process capability studies 15 times like American Auto. They just say, "take a bit of material off here and there and that will be okay—lets go." In 11 years, I have never built a prototype tool for Toyota. Knee bolsters, floor panels, IPs, etc. are so similar to the last one it is not necessary to build a prototype. When there is a problem, they look at the problem and come up with a solution—focus on making it better not placing blame (Liker, 2004).

Toyota has turned its structured "bureaucratic" relationship with suppliers from coercive to enabling (Adler, 1999) with stable processes and clear expectations. In contrast, most U.S. automotive companies constantly reinvent their supplier processes, which can vary from engineer to engineer and even department to department within the same project. Without developing a fair and stable business base, it is impossible to get to the higher levels of enabling systems and truly learning together as an enterprise (Liker, 2004).

There is another significant disadvantage to the U.S. auto companies' highly bureaucratic, competitive bidding approach to supplier relationships: the tremendous hidden factory and transaction waste associated with the maintenance of this system. Searching the globe for the lowest cost means managing very large numbers of suppliers as well as introducing a steady stream of new suppliers into your system. These suppliers are unfamiliar with your requirements and demand a great deal of attention to get up and running. While administering complex contracts, managing global bidding wars, and overseeing the constant introduction of new suppliers into the process, U.S. automakers must maintain mammoth purchasing organizations, deal with incredibly cumbersome and slow sourcing processes, and live with constant variation of supplier performance in the development process.

The bottom line in a lean PD process is that you treat suppliers respectfully and reasonably. After all, they are engineering and building critical portions of the product that you are trying to sell to the customers who you want to view the product as world-class goods. If the supplied parts are inferior, the product you sell will be inferior. If the suppliers are inferior, you should stop using them. However, if the supplier is engineering and building world-class components, you should treat that supplier with respect. In Henke's survey of auto suppliers, trust was the most important measure ranked, and Toyota consistently scores at the top of automotive companies. Without trust, there is no partnership.

Trust between Toyota and its suppliers is a two-way street; each partner benefits because each partner is willing to go the extra mile down this street to keep the relationship strong. As this chapter has illustrated, Toyota demands much of its suppliers. During its global CCC21 initiative, for example, Toyota asked its suppliers to reduce the price charged Toyota by 30 percent for the next new model. This sounds like an impossible goal, especially given suppliers' tight profit margins. But Toyota never just dumps a demand on suppliers; it makes a request and then works with the supplier to achieve what is needed. In this instance, if Toyota had asked for a 5 percent reduction, the supplier might have simply taken it out of profit margin. Because taking out 30 percent would be prohibitive, the supplier must find a different solution and this means examining every aspect of the business from design to raw material to delivery to find true cost savings. As Darrel Sterzinger, general manager of engineering design chassis, at Toyota Technical Center explained:

> A true North American supplier cannot imagine 30 percent—it boggles the mind. But when I sit down with them and explain Toyota's thinking, they can understand the purpose of it. It is not the 30 percent we are thinking about. It is a new way of doing business. We explain to them and look at their operation and then they feel more comfortable. They think we are like the big-3 and Toyota has gone wacko. If you just tell the supplier 30 percent without working with the supplier—it is unreasonable. But for us it is a whole way of thinking which starts with the design. If we do not think about working together with the supplier as a partnership and start talking about cost competition with other suppliers, like GM, the suppliers will be very nervous.

Toyota never bullies or threatens its suppliers, which may explain why it first termed the Toyota Production System the "respect for humanity system." This philosophy extends to all the partners with which Toyota does business. The bar is set high for the supplier, and the consequences of failing to step up and improve to Toyota's standards can be painful. However, there are financial rewards for working hard to improve as well as the pride and satisfaction that come from success and recognition.

The LPDS principle to *fully integrates suppliers into the product development system*, means you must first get your own house in order, then go out and find only the most qualified companies to partner with, working with them to develop compatible systems and procedures. Ultimately, this leads to working and learning together as an enterprise, which leads us to the ninth LPDS principle: *Build in Learning and Continuous Improvement.*

LPDS Basics for Principle Eight

Fully integrate suppliers into the product development system

Customers view purchased components as a part of the product and want the whole product to be problem free. They do not care if a problem was not your fault because your supplier was not reliable. Whoever sells the final product is responsible. For this reason, suppliers of core components must have the same level of engineering and manufacturing capability to contribute quality parts as your lean enterprise has in engineering and manufacturing quality products. In addition, suppliers must be compatible. They must fit seamlessly into your product development system, your launch system, and your manufacturing system. This requires learning how to work together through repeated experiences. It is your responsibility to communicate clear expectations. To accomplish this, Toyota brings selected suppliers into the simultaneous engineering process very early in the concept stage. Suppliers make a serious contribution to simultaneous engineering, knowing that they are investing ahead of the payback that will come in the production stage. This is something to emulate.

CHAPTER 11

Build in Learning and Continuous Improvement

*The ability to learn faster than your competitors may be
the only truly sustainable competitive advantage.*

DE GEUS (1988)

AT TOYOTA, LEARNING AND CONTINUOUS IMPROVEMENT are fundamental to
how each person does his or her job every day and are inseparable from
Toyota's culture. Toyota sets challenging performance goals for every project and holds both real-time and postmortem learning events (called *Hansei* or reflection) that encourage functional specialists to verify and update
their own knowledge databases. Learning and continuous improvement
are also embodied in a problem-solving process that develops root-cause
countermeasures: multiple potential solutions that prevent recurrence
(Ward et al, 1995). LPDS Principle 9 postulates that learning mechanisms
are embedded into the product development process; that learning excellence is a basic characteristic inherent to a true lean learning organization;
and that effective learning and continuous improvement may be the most
sustainable competitive weapons in its arsenal. In fact, Toyota's awesome
ability to learn quickly and improve at a regular cadence may well be the
characteristic of Toyota its competitors should fear most.

Defining Knowledge and Organizational Learning

Many theories purport to define organizational learning, knowledge
transfer, information management, and how these apply to product development. It can be argued, in fact, that knowledge and information are the
stock and trade of product development. Literature on knowledge and
learning also provides numerous views on organizational learning and the
requirements for a successful learning organization. Senge (1990), for
example, claims that organizational learning is the "ability of a group of
people to consistently create the results the members truly desire." Peter
Drucker (1998) observes that the future belongs to the "knowledge-based

enterprise" and, in his book, *The Knowledge Creating Company* (1995), Ikujiro Nonaka concludes that the "one source of lasting competitive advantage is knowledge." Some scholars go a step further, maintaining that the purpose of an enterprise is creating, storing, and applying knowledge (Kogut & Zander, 1992; Conner and Prahalad, 1996).

In his article "Teaching Smart People How to Learn" (1998), Chris Argyris argues, "Corporate success depends on learning" and postulates two types of learning: *single loop learning* and *double loop learning.* Single loop learning comprises error detection and correction without change to the underlying values of the system. This can be compared to a thermostat detecting a decrease in temperature and turning on the heat without the ability to question what it is doing. Double loop learning occurs when the actual operating norms are questioned, which is fundamental to an organization's ability to learn.

Other scholars point out two very different types of knowledge: the first is *explicit knowledge* or information easily codified and transferred without significant loss of content. Explicit knowledge includes "facts, axiomatic propositions, and symbols" (Kogut & Zander, 1992). The second category of knowledge is *tacit knowledge* or know-how. This type of knowledge is complex, hard to codify, and difficult to transfer; it requires dense ties and long relationships (Nelson & Winter, 1982; Kogut & Zander, 1992).

David Garvin (2000) affirms, "Knowledge is generally seen as a key corporate asset to be leveraged and exploited" and defines a three-step process required for organizational learning: information acquisition, information processing, and application. Argyris (1992) emphasizes that the litmus test of learning requires action: "From our perspective, therefore, learning may not be said to have occurred if someone discovers a new problem or invents a new solution to a problem. Learning occurs when the invented solution is actually produced." Jeffrey Pfeffer and Robert Sutton echo this sentiment in their book *The Knowing-Doing Gap* (2000), in which they assert that acting on knowledge is the fundamental difference between successful and unsuccessful learning within organizations. In fact, Pfeffer and Sutton maintain that for all the attention given to organizational learning, little is accomplished. Garvin agrees: "Learning organizations have been embraced in theory but are still surprisingly rare."

Explicit Versus Tacit Knowledge Transfer

Most companies focus on explicit knowledge, defined above as easily codified, transferred without significant loss of integrity, and generally stored as facts, axiomatic propositions, or symbols. Historical dates, mathematical equations, and formulas fall into this category. Explicit knowledge is sometimes referred to as "know what" knowledge. It is characterized by voluminous databases.

By contrast, tacit knowledge is complex, "sticky," and difficult to transfer. Sharing tacit knowledge requires intricate ties between participants; it entails longer, deeper relationships, such as those that develop between a master craftsman and his apprentice. In fact, the apprenticeship tradition was designed as a means to transfer tacit "know-how" knowledge from master to student.

Dyer and Nobeoka (1998) suggest that tacit knowledge holds the most competitive potential for companies even though it is difficult to learn (you cannot merely imitate it), manage, and apply. Because it makes effective organizational learning difficult, many companies prefer to focus on explicit knowledge, which can be more easily gathered and stored. The big problem with explicit knowledge is that it can also be imitated. If one company can create an extensive database of explicit knowledge, so can a competing company, and this dilutes the competitive advantage of both.

One of the main reasons that companies fail at imitating lean systems is that they mistakenly copy only the "explicit" knowledge of lean tools and techniques. By and large, these companies attempt to implement lean without understanding the need to tap into the tacit knowledge of lean culture, the know-how knowledge that enables an organization to learn organically, adapt, and grow. They fail to grasp that *in highly technical environments, such as product development, tacit knowledge is the true source of competitive advantage.* One important study on automotive product development supports this assertion (Hann, 1999). The study revealed that die tryout knowledge is highly tacit, tends to be part specific, and is very difficult to master. The author of the study also found that specialization, strong work routines, and continuous work fostered a significant reduction in time to complete die tryouts. In combination, these findings loosely define the power of effective lean learning.

Toyota's Product Development Learning Network

From its inception, Toyota has consistently worked on developing ways to collect, disseminate, and apply tacit knowledge, creating a learning network that is applied enterprise-wide from product development to manufacturing. To understand this is to understand a piece of Toyota's culture, with its strong emphasis on the importance of people. For Toyota, transferring tacit knowledge means that people must get to know each other well enough to share deep insights. Most often, this happens face to face and one to one. What enables this transfer is a shared social philosophy, which is consistently nurtured and sustained. There are several ways Toyota forms a highly effective learning network in product development.

1. *Supplier technology demonstrations.* At the beginning of each program, suppliers demonstrate technology that might be appropriate for the new vehicle. This is a good opportunity for Toyota engineers to learn about new developments (Kamath & Liker, 1994) and for Toyota to leverage supplier resources fully. Toyota suppliers bring parts and meet face to face with Toyota engineers.

2. *Competitor teardown analysis.* Teardown exercises provide an opportunity to learn about competitors. These hands-on exercises are another example of *genchi genbutsu* and an excellent way for engineers to learn. In lean PD, you also want to use these learning tools to disseminate information so that all team members learn from them. (Chapter 15 elaborates on this.)

3. *Checklists and quality matrices.* Toyota uses these tools to organize and store information so that people can apply it similarly across the organization. (Chapter 15 elaborates on this.)

4. *Learning focused problem solving.* This represents the A3 problem-solving discipline that helps people learn while seeking lasting solutions. At Toyota, problem solving begins early (at the source), is data driven, and includes a learning component. (Discussed in Chapter 14.)

5. *Know-how database.* The know-how database is the collection of standards combined with design data and tools, such as digital assembly (discussed in Chapter 15). The functional organizations that use these databases (evolved from checklists) maintain, validate, and update them as needed.

6. *Hansei events. Hansei* is a Japanese word for reflection. At these reflection events, participants share their PD program experiences, lessons learned, project shortcomings, and then discuss and develop countermeasures. (*Hansei* events are discussed in more detail later in this chapter.)

7. *Program manager conferences.* Program managers from various projects meet once a year to discuss lessons learned and to pass

on new standards. The lessons are derived from each program's *hansei* or reflection events that the program manager leads.

8. *Business Revolution Teams.* These cross-functional teams are formed to address single revolutionary improvement efforts. Toyota high-level executives assign people to these teams full time and each assignment can last six months to a year. The first hybrid, *Prius*, began with a business revolution team, as did Toyota's effort to eliminate engineering changes.

9. *OJT skills matrices and learning-focused career paths.* Specific OJT skill matrices and mentoring technical skills teach engineers leadership and sets their career paths. Learning is built into everyday assignments so engineers can use Toyota's scientific method—identify, evaluate, countermeasure, verify, communicate (standardize)—to question and learn continuously. This cycle of mentoring, learning, and applying until one has the potential to mentor others is the heart of a lean learning system.

10. *Resident Engineers (RE).* Engineers are exchanged on temporary assignment both within Toyota and with affiliated companies. This is a learning opportunity for the RE and is also a method of standardizing practices.

Learning from Experience

If there is no mechanism or culture in an organization for capturing, retaining, and reusing knowledge, that organization is almost certainly engaged in constantly reinventing the wheel. That said, it is only fair to admit that learning from experience is often easier said than done, particularly when it involves very complex, episodic experiences. The process requires conscious reflection, and very few organizations do this well or even see the importance of doing it. According to Garvin (2000), there are a number of reasons the learning from experience process fails.

1. *Time pressures.* Stressful time pressures impact most people in every organization. As soon as one project is over, another one (sometimes already behind schedule) looms and must be immediately addressed.

2. *Oppressive workloads.* People tend to be absorbed by their current workload, often ignoring important things learned from past (even recent past) experience.

3. *Blaming.* Organizational lessons-learned events often turn into sessions marked by scapegoating, finger-pointing, and assigning blame. As a result, only the loudest or most powerful are heard at such events; other people are reluctant (and often silent) participants.

4. *Complex projects.* Complicated projects thwart understanding, causing frustration.

Garvin goes on to identify reflection, *hansei* in Japanese, as one of the most potent learning mechanisms to confront these four pitfalls and has five suggestions to help reflection events succeed.

1. *Hold the reflection as soon as possible after the event.* The loss of valuable real-time information increases over time.

2. *Focus on things under the group's control.* People should not waste time confessing other people's sins or policies that the group cannot take action on.

3. *Tolerate criticism and honest dialog.* Participants should feel free to express all relevant views without fear of repercussion. Personal attacks should not be tolerated.

4. *Must be a regular event.* People must see this activity as part of the job.

5. *Must update standards and processes.* Just talking about it doesn't fix it. The end result of each event must be some concrete, observable improvement or people will view these events as worthless and stop participating.

Hansei at Toyota

Hansei (reflection) has deep roots in the Japanese culture and begins in child hood as a sort of time out when parents ask their children to reflect on some juvenile indiscretion. *Hansei* requires a certain humbleness or egolessness in order to search for and identify weakness in ones performance or character. It can be a very difficult process especially for western engineers who often find the process quite foreign and whose egos can sometimes get in the way. *Hansei* can sometimes be seen by westerners as overly negative because of its focus on weakness since Japanese engineers see little reason to discuss things that were done well. But *hansei* is a necessary and powerful process for continuous improvement. In fact George Yamashinta, former president of TTC says that in engineering there can be no *kaizen* without *hansei*, as the nature of engineering work makes it necessary to think deeply about the process. Each Toyota *hansei* event or *hansei kai* (reflection meeting) is designed to enhance organizational learning from experience. There are three types of *hansei*:

1. *Personal reflection.* In this *hansei* exercise, a supervisor asks an engineer to reflect on some aspect of his or her performance and develop an action plan for improvement. A written response is expected and will be reviewed by both parties. Specific goals are set and a follow-up plan is outlined. This activity addresses a specific skill or capability that needs to be improved.

2. *Real-time reflection.* This *hansei* experience occurs at a group or team level and is scheduled into the product development process. These *hansei kai* usually take place at major milestone events, suchas final data release or tool transfer to the stamping plant. PD programs can last a year or more and you can lose much information in that time, so these are *regular events* scheduled to take place as soon as possible after the actual activities. The *hansei* event can focus on a specific issue, such as looking at the problem of some parts that arrived late for the prototyping event. They might also reflect holistically on the prototyping event. Depending on the issues involved, group or team *hansei* events can be cross-functional or intra-functional. They are most often cross-functional, focusing on internal customer feedback. Although flexible, they typically follow a general outline and address the following questions:

   a. What were our goals and objectives? (often organized by category)
   b. How did we actually perform to our goals?
   c. Why did it happen? (data analysis)
   d. How will we improve next time? (a written improvement plan)

   These *hansei* events often lead to updating a relevant standard or may require an A3 for reporting or problem solving. The length of *hansei* events can vary significantly, but they are often scheduled as two or three meetings, each lasting two to four hours, within a two-week time period. Much of the actual "prep" work occurs before the meeting date. Toyota's A3 and "5 why" problem-solving methodologies are rigorously employed.

3. *Postmortem reflection.* This is the "what went right, what went wrong" lessons-learned event. The presentations are formal, with most of the real analysis and reflection occurring before the meeting date. At the meeting, representatives from each of the functional groups review program performance and discuss results as well as new ideas stemming from real-time reflection. Program managers talk to participants to get feedback and synthesize these reflections from their programs into a small number of lessons learned that are shared at PM meetings held several times a year. At these meetings or *hansei kai,* program managers produce written summary documents that are shared within and across other PD program teams. Note that there is a distinction between the program manager and the chief engineer and that these roles are not interchangeable. The chief engineer is responsible for the product. The program manager (not as high a level) is responsible for the process.

Ijiwaru Testing at Toyota

Testing and validation can be another important opportunity to learn from experience in product development. At NAC, required part performance specifications are set in advance and designs are tested for compliance to these specifications. Learning in this environment is minimal because it is strictly a pass-fail metric. *Ijiwaru* testing on the other hand is the practice of testing subsystems to failure. By testing these subsystems under both normal and abnormal conditions and pushing designs to the point of failure Toyota engineers gain a great deal of insight into both current and future designs and materials by understanding the absolute physical limitations of their subsystems. Consequently, Toyota engineers develop a depth of product knowledge that is exceptional in the industry. This practice also gives Toyota a great deal of confidence in the performance parameters of their products in the hands of their customers.

The Power of Problems

It should be pretty clear by this point that Toyota views problems as a natural part of product development. In fact, in a sense, the essence of product development can be seen as a series of technical problems that must be identified and solved. From this perspective the companies who excel at technical problem solving will do well at product development. And companies that improve their technical problem-solving capability will improve their product development competence. Consequently, Toyota looks for and values technical problems as opportunities to learn, grow, and improve its performance and encourages positive problem-solving behaviors. Confronting, solving, and learning from problems is one of Toyota's greatest strengths as a product development company. Taking this approach will help companies to confront and solve problems early, arrive at optimum solutions quickly, and provide permanent resolution. This increases a company's knowledge base, builds critical skills, and introduces superior solutions that become a permanent competitive advantage other companies will find difficult to replicate. In short, the lean PD process views problems as opportunities—not only to improve products but also to improve the core of a company's product development capability.

By contrast, at NAC it is common to see problems as negative and unexpected, an attitude that suggests problems shouldn't occur at all. When problems do surface, as they inevitably must, there is a lot of finger-pointing or blame-gaming. This happens because managers and workers at these companies see problems as indicators of poor performance. As a result, individual engineers learn quickly to hide problems for as long as possible. Of course, the hidden problems fester, and as the product design continues to progress into a later cycle, they become even more difficult to resolve. Once the problem is discovered, people have to scramble and backtrack; other work is put on hold until the fire drill ends.

The Japanese Toyota engineering supervisors at the Toyota Technical Center in Ann Arbor tell an interesting story about some American engineers they hired into the center from competitor companies. As the supervisors made their rounds and spoke to these engineers, they routinely asked if they had any problems. The engineers automatically responded, "No, no problem." This response, *mondani* in Japanese, soon became a semicomical in-house joke in which *mondani* was transformed into "Monday night!" Although it began as nothing more than an amusing expression, it turned out to be a very unamusing source of concern. When the first design review forced problems into the open, Japanese supervisors quickly learned that "no problem" was a big problem indeed.

Problem Solving at the Source

The prototype phase of product development is a period of intensive learning. As part designs become prototype parts, technicians and engineers begin to assemble them, often encountering problems that they must resolve in real time. In an environment fraught with problems, it is clearly advantageous for a company to achieve high-quality, permanent solutions (or countermeasures) quickly. To develop and sustain a lean PD process, a company will need to transform these problems into organizational learning opportunities and subsequent continuous improvement. This means investing considerable time and resources into perfecting problem-solving capability so that there are mechanisms for capturing, verifying, codifying, and sharing these solutions *before they are lost*. Toyota accomplishes this in a number of ways, beginning with the standardized problem solving during *kentou*, which continues throughout the PD process. The standardized scientific problem-solving process:

- identifies the problem's root cause
- evaluates the potential impact of several possible solutions
- produces a high quality countermeasure that can resolve the immediate issue as well as prevent its recurrence
- verifies the countermeasure during subsequent *hansei* events
- communicates the result across programs by updating standards and checklists, which are increasingly part of the "know-how database."

There has been a great deal written about Dr. Deming's Plan-Do-Check-Act (PDCA) process and Toyota's practice of asking why five times to solve problems at the root cause and we will not elaborate on it here. However, it is important to note that those methods are equally applicable in product development as the shop floor. In product development it is crucial to solve problems early, at the source, and permanently; and to learn from these problems in order to improve the organization. In a lean PD process, it is essential to build solutions into the continuous improvement process; these become the starting point (the standards) for the next program.

Cross-checking

Cross-checking is one method employed by Toyota to discover problems and check quality, especially from the prototype phase onward. This applies to the process of understanding the true condition of parts and the appropriateness and accuracy of the measurement system that is being employed. You can achieve cross-checking by requiring several groups to check the same parts/data independently. For example, prototype parts might be checked on a supplier's gauge, checked to templates by the body engineer, and finally checked to a fixture at the build by a prototype technician. If someone identifies a discrepancy, rather than having one group necessarily override the other, a detailed investigation is initiated to find the root cause. In many cases, it is a measurement error (part locating, holding, and gauging). If this is the case, the team will incorporate a subsequent countermeasure on production gauging and fixtures and, if appropriate, will update the standards.

Daily wrap-up meetings

Another potent learning and problem-solving mechanism utilized during design reviews, prototype builds (physical and virtual), tool manufacture, and launch is the daily wrap-up meeting. Held at the end of each day, typically on the shopfloor where the work is being done, the wrap-up meeting is attended by all key participants, including suppliers, and is a strategy that captures lessons learned, clarifies assignments, and generally aids in real-time, course-correction decisions.

Ignorance: The Ultimate Expense

Many companies view paying for employees to learn as an unnecessary cost. Anything that is not tactical or does not produce immediate, measurable results is "fluff," and those who endorse an opposing view are not considered serious-minded, action-oriented business people. But the absence of deep technical understanding drives a "more is better" philosophy, which leads to more elaborate gauges, more reviews, more audits, more inspections, and more checkpoints (because it is always "safe" to check and check again).

This is an incredibly expensive approach to product development. Companies that operate from this perspective heap quality initiative upon quality initiative (QS 9000, ISO, QOS, Shanin, Six Sigma) and commit scarce human and financial resources to these add-on initiatives, at the expense of core engineering capability and real towering technical competence. On the surface, these quality initiatives demonstrate a "commitment to quality" and make people feel that they have accomplished something. In truth, however, they fail miserably in achieving those things that guide an intelligent and balanced approach to fostering quality in a PD process: the time, dedication, and hard work required for a deep technical understanding of that process.

Such companies, rather than understanding what is truly critical to vehicle quality, generally focus on the trappings of quality rather than the essence of quality. As Deming noted, however, you cannot "inspect in quality"—and this applies to product development and manufacturing. Furthermore, constant inspection tends to create a culture of fear. Engineers with a "common sense" technical understanding of the critical essentials, are often afraid to speak out in this kind of environment, because they risk being labeled "anti-quality." For this reason alone, the "quality trappings syndrome" is an insidious and potentially deadly resource-sucking spiral that can drain the life out of a product development system.

Rapid Learning Cycles

Learning from experience is enhanced by repetition, and this is especially true for complex or difficult tasks. In automotive product development, new engineers acquire experience during multiple vehicle development programs to master their discipline. In companies with very slow-moving development programs and frequent job rotation, engineers rarely have the opportunity to experience more than one development program and focus only on one aspect of the product. On the other hand, companies with very fast-moving product development programs and less frequent job rotation have many more learning cycles for engineers across multiple products—resulting in much shorter learning curves.

Once again Toyota takes Deming's analysis of the Shewhart PDCA cycle—plan, do, check, act—very seriously. Every project progresses through this cycle. Within a product development program, each major phase is a mini-cycle of the PDCA model; the entire PD program is a macro-level reflection of the cycle. The faster the product development cycle, the more can be cycled through it. Most importantly, PDCA develops towering technical competence and supports continuous learning, and *Build in Learning and Continuous Improvement* may be the most important LPDS principles for companies beginning to develop a lean PD system. The next chapter connects this principle with the final LPDS principle of the people subsystem.

LPDS Basics for Principle Nine

Build in learning and continuous improvement

The ability to learn and continuously improve may be a lean product development system's most powerful competitive weapon. Within this framework, it is the tacit, "know how" knowledge that is most potent, the most difficult knowledge to foster and manage. There are no short cuts or IT solutions. Transfer and application of tacit knowledge requires close ties and a good deal of time of the sort enabled by structured mentoring, OJT, strategic and aligned training, and lots and lots of face-to-face time at the source. Toyota understands the essentials of its business and builds tacit learning into the way work is performed every day. At Toyota, tacit learning is not an add-on or extracurricular activity. It is core curriculum that is learned through *hansei*, mentoring, PDCA cycles, excellence in problem solving, teardowns, and other experiences, all of which are focused on improvement.

CHAPTER 12

Build a Culture to Support Excellence and Relentless Improvement

TPDS is rooted much deeper in the culture in things like genchi genbutsu, *the chief engineer system,* kaizen, *TPS, etc. It is the totality of it working together in the culture established across many years that makes it all work. What is actually happening in your workplace? A good understanding of that is critical. To have a clear understanding of what your work is and how you are doing, that is what is important.*

TAKESHI UCHIYAMADA, Chief Engineer of the original *Prius*

AN ORGANIZATION'S CULTURE DEFINES what goes on in its workplace, and no company can develop a lean PD system without a strong and vibrant culture. This chapter, which examines LPDS Principle 10, highlights a number of significant cultural elements at Toyota. All the other principles discussed to this point are viable because culture makes them a living part of how work gets done in a lean environment.

How Culture Can Stand Between You and Lean

Companies that have achieved impressive results applying lean concepts to manufacturing often think of applying lean concepts to product development. One reason for this is waste. As noted in Chapter 5, waste exists in both manufacturing and product development. The assumption is that if lean has eliminated or reduced waste in manufacturing, it can do the same for product development. To assist this transformation, company engineers attend courses on lean product development. Invariably, their organizations expect them to bring back effective waste-busting tools that will cut lead-time and cost. Of course, it is not that simple. Comments from companies whose engineers have taken such courses or have attempted to apply Toyota's lean PD tools illustrate the problem:

- We invested millions in a "book of knowledge." It is a web-based system and several people were assigned full time to load it up with standards and best practices. Best practices were solicited from the field and entered into the system by someone in IT. But we are getting almost no hits—engineers are not using it.
- We created a new role of chief engineer. A bunch of engineers in different project manager roles were given this new title. But they still acted just like the old project managers and still did not have any power to get anything done.
- We value stream mapped and came up with great ideas. We created A3s and developed action plans. Then we got three new programs dumped on us, the crisis mode kicked in, and the action plans went out the window.

As these examples show, companies cannot simply have their engineers learn and apply these powerful tools and then sit back and watch the waste evaporate. What is missing is a lean culture to sustain the tools. Loosely defined, culture is the soft, imprecise, fuzzy stuff of everyday life. Within any company, it is what people think and believe and what drives daily priorities. Leadership and a company's culture are inextricably intertwined. Culture defines who emerges as a leader and leaders define the culture. Edgar Schein (1988), one of the leading authorities on organizational culture, provides a more formal definition of culture as:

> ... the pattern of *basic assumptions* that a given group has *invented, discovered,* or *developed* in learning to cope with its problems of external adaptation and internal integration, and that *have worked* well enough to be considered valid, and, therefore, to be *taught to new members* as the correct way to perceive, think and feel in relation to those problems.

An explication of four key concepts in this definition follows:

1. *Culture operates at an unconscious level.* By basic *assumptions,* Schein means that culture, our core belief system, begins at an unconscious level, the roots of which go back to early life experiences. As individuals grow up and mature, they learn what is right and wrong, what is good and bad, what they enjoy and what they want to avoid. In much the same way, as people enter an organization and begin to find their way around that organization, they

learn what it values and rewards and what it punishes. People modify their behavior accordingly, even though their deep-seated personal cultural assumptions remain. The Toyota culture uses many basic assumptions for defining the best way to "perceive, think, and feel." This is especially true in relation to problem solving and is a major reason why Toyota's PD system is difficult to teach—woven throughout its many tools and techniques are assumptions on behavior that are rooted in both the Toyota and Japanese culture. For those who grew up in Japan and who have been at Toyota for many years, it is difficult to articulate a culture that, for them, is second nature.

2. *Necessity-driven and empirically-based production system.* Toyota's PD system was *invented, discovered,* and evolved over decades to cope with Toyota's unique challenges in dealing with external adaptation and internal integration. Toyota's external adaptation challenges started with building an automotive company from scratch in a weak, post-World War II economy. The challenges of internal integration were shaped by the "collectivist" Japanese environment in which most companies assume employees should subordinate their individual desires to the needs of the company that employs them. This cultural and economic context—the fact that the people had to band together to survive—made it relatively easy for Toyota to get suppliers to buy into its specific goals and processes. A corollary to this is that a collaborative spirit is essential to lean product development. Historically, Toyota's PD system emerged from trial and error and from a scientific method used to find real solutions to real problems that were rooted in the broader socioeconomic context of the time.

3. *Adapting systems.* Toyota has made a tradition of adapting new methods that work and fit into its cultural framework. Very simply, if a system or tool works, it is kept and used but only after it is adapted precisely to a Toyota process. If it cannot be adapted, it is dropped. This reflects Toyota's empirical approach: working to find real solutions to resolve real problems in actual experience and not falling into the bureaucratic trap of letting internal beliefs or policies dictate the best way of doing things.

4. *Learn the system by doing.* Toyota uses an explicit approach for teaching the Toyota PD system to new employees in real-time and real situations. People do not learn the Toyota Way in a classroom

or online. As discussed in Chapter 9, senior mentors take on the responsibility of developing subordinates, and young engineers learn by doing. Learning at Toyota is also a process of socialization. Although leadership style is not identical, every Toyota leader teaches the same core assumptions with a very clear set of beliefs about the philosophy of the Toyota PD system. The expectation is that everyone will have the same shared beliefs and clear precepts about what is considered good and bad practice in engineering and work toward the same goals.

Things that support how to "perceive, think, and feel" in relation to problems include some concepts already covered in this book: *genchi genbutsu*, set-based thinking, reverence for technical excellence, *hansei*, and putting the customer first. This broadly-shared cultural DNA is fundamental to the success of lean thinking. Paradoxically, this same cultural DNA is one of the reasons it is a challenge, even within Toyota, to teach the lean product development system to new employees globally.

In his comprehensive explanation of culture, Schein also explores the "strength" of a culture. By definition, culture is shared among members or a group. Of course, not every individual will perceive, think, and feel in the exact same way. In statistical terms, the shared variance is the culture, and individual variation indicates deviations from the culture. Thus, a strong culture is one in which many things are shared by a large cross-section of people. Of course there are good and bad strong cultures, highly effective and highly ineffective. Obviously, 100 percent consistency is impossible to achieve—people, after all, come from different backgrounds. Diversity of views can be a strength in problem solving. But Toyota works hard to bridge differences, and as a result, most Toyota employees do share basic assumptions about values, priorities, and how to get work done. In other words, culture at Toyota is strong.

A Tool Is Not a Solution

An engineering vice president who heard about Toyota's A3 reports required every engineer in his organization to develop at least one A3 per quarter. Engineers took their original reports and spent days squeezing them down to A3 size to meet the requirement. But simply buying reams of A3 paper, requiring engineers to write A3 reports, and expecting to get the same results as Toyota will not work. At Toyota, A3 works because it takes place within a particular cultural context (see Chapter 14). As the reader has already seen, Toyota has a long history of training engineers in rigorous methods to collect information, identify real problems, get at critical essentials, consider multiple potential solutions, get broad input through *nemawashi,* creatively develop root cause countermeasures, and use discipline in implementing. Once all of this is in place, Toyota engineers follow up with standardization and continuous improvement. A well-socialized Toyota engineer almost intuitively knows that getting input from all the right people is necessary and he or she has learned from experience the value of presenting information concisely and visually. Without this cultural context, the A3 is a mechanical requirement, an exercise in summarizing complex information to please the boss. Without a deep thought and consensus-based process, the A3 report is a tool generated for the wrong reasons. In lean thinking, *it is the process of producing not the result of producing the A3 that makes it a powerful method.*

Contributing to Customers, Society, and Community

As described in *The Toyota Way,* the roots of Toyota's culture can be traced to the founder of Toyoda Automatic Loom Works, Sakichi Toyoda. Established in 1926, the company was Sakichi Toyoda's response to seeing his mother and others working their fingers to the bone weaving on primitive manual looms. The founder's invention eventually led to patents, globally respected Toyoda power looms, and great wealth. Sakichi Toyoda could have retired a rich man and let his son inherit his wealth and live in luxury. But the purpose of the loom company was to contribute to society, and to fulfill this purpose, so Sakichi Toyoda challenged his son, Kiichiro, to continue this spirit of contribution. When Kiichiro returned from England after selling a loom license to the Platt Brothers for 100,000 pounds, he used these funds to finance the Toyota Motor Company, which was established to continue the family's legacy of contributing to society.

There are several ways Toyota pursues this idea of social contribution. "Customer first," for example, is a basic belief and a lean principle that a company should exist to serve society. In many companies, a customer-first attitude is replaced or subverted by a me-first culture, which trickles down from the executive suite all the way to the engineers and the workers on the shopfloor. Another way Toyota contributes to society is by providing jobs in communities where Toyota cars are sold. A company that focuses as much attention as Toyota does on eliminating waste in manufacturing, might find it pragmatic to lay off employees during slow periods. Toyota's practice is to keep employees on the job, relying on attrition or severance incentives that encourage employees to leave voluntarily rather than laying them off. As the LPDS Principles 5 through ten show, Toyota is about people and for people. When Toyota managers say, "people are its most important resource," it is not a hollow platitude. To this day, the Toyota Way is to think beyond individual and short-term concerns to the long-term good of the company (including the employees of the company) and society.

Technical and Engineering Excellence Are Intertwined in the Culture

As discussed in Chapter 9, for most companies, technical excellence is about hiring people with the right skills and educational background. During the hiring process, a company may make an effort to determine whether a candidate will "fit" in the company environment. Once a candidate is hired, however, there is often little (if any) effort to instill or reinforce deep-seated cultural values. This is counterproductive. In a lean PD system, getting a top student from a top engineering school guarantees only one thing—a smart young person who has an opportunity to learn to be a real engineer. At Toyota, formal education is a foundation that is not, in and of itself, useful until employees have been fully acculturated. Toyota's cultural structure includes the following precepts:

- Run by engineers, making it a technical hierarchy
- Fundamentally a manufacturing company that translates into core value-added manufacturing activities and those supporting manufacturing
- Focuses on developing technical mastery using the "hands on" Toyota scientific method
- The individual engineer is central to the product development system
- Learning and continuous improvement (every-day *kaizen*) is fundamental to how work gets done
- Process discipline, hard work and loyalty is expected of everyone (especially leaders)
- Data focused

These cultural precepts go back to Sakichi Toyoda, the "king of inventors," who learned from the ground up how to build power looms that used steam engine technology. Following his father's example, Kiichiro

Toyoda likewise learned basic automotive technologies and processes from the ground up. Given this tradition, some might find it strange that Fujio Cho, who became president of Toyota Motor Company in 1999, was not from the Toyoda family or even an engineer but a lawyer who graduated from the University of Tokyo in 1960. How, one might ask, could a lawyer run a technically-oriented manufacturing company? The answer lies in the way that Cho was acculturated.

After joining Toyota, Cho spent 14 years in different staff jobs. Then in 1974, he was assigned to work with Taiichi Ohno to apply TPS to administrative operations. From that point on Cho's life changed. He was mentored by Ohno, whose style was a harsh combination of beating up, pressuring, and cajoling. Under Ohno's guidance, Cho learned that to improve administrative operations he had to go back to the basics of understanding how TPS applies in the factory. Schooled in the harsh college of Ohno's TPS—all on the shopfloor—Cho eventually became an expert in TPS. He was so respected within Toyota that in 1988, he was assigned to open Toyota's first wholly-owned factory in the United States, the Georgetown, Kentucky, plant. At the plant, Cho made daily visits to the shopfloor to teach TPS directly to shopfloor team members and leaders. His immersion in technical practices on the shopfloor later earned Cho an honorary Doctor of Engineering degree from University of Kentucky. It was the sum of these experiences that ultimately led to Cho being named president of Toyota Motor Company, a position that he was able to achieve only with a deep understanding of TPS.

The point of this story is twofold. First, it illustrates how Toyota develops, mentors, and invests in its personnel. Secondly, it illustrates what a culture of technical excellence means: Everyone in the company, from the shopfloor worker to the president, believes that the purpose of the company is to add value and that the way to add value is by perfecting a technical process. In lean thinking, having people do technical work to perfect technical processes is the highest value within a company. Nontechnical personnel exist to support this primary mission.

As emphasized throughout this book, Toyota is a manufacturing company run by engineers. This is important in a lean PD system because it makes the individual engineer the center of that system—all decisions, processes, and ideas germinate from the individual engineer working in teams and reporting up to the chief engineer. What develops during this process is "towering technical competence" that comes from learning by doing from the ground up rather than from technical training exercises that "upgrade" engineers. In a lean PD system, being an engineer is a calling rather than a job. In combination, these are the elements that *make a company a culture of technical excellence* that is firmly grounded in technical mastery, both in product development and manufacturing.

Discipline and Work Ethic

Technical excellence can mean many things. There is, for example, the image of the brilliant but absent-minded inventor who gets intensely passionate about the invention of the moment, losing track of time and place. The brilliant inventor can be completely disorganized, go on work binges, make a mess and then take time off, leaving other people to try to clean up the mess. Turn 180 degrees and you see what technical excellence means within the Toyota culture, which expects something quite different. Toyota engineers, like the brilliant but absent-minded inventor described above, are expected to work hard, work long hours, demonstrate significant technical knowledge, and be passionate about their work. However, the key difference is that they must be dedicated to The Toyota Way and committed to the discipline that entails.

When Toyota developed the first *Prius*, then President Okuda proclaimed that the car was the future of Toyota. Toyota engineers worked slavishly, canceling all vacations, to meet timelines for a car that was, in many respects, a total novelty in the automotive industry, one that required new and unfamiliar technology. In September 1996, after years of working on the concept, the engineers involved in this preliminary process formally presented their styling concept to the board for program funding approval. The board approved the detailed plan and funding for development and what ensued was a marathon race to reach President Okuda's target date of December 1997—an unprecedented 15-month development cycle (which later moved even earlier to October). Everyone understood that achieving this aggressive goal and its corresponding timing targets meant personal sacrifices. As an example, Mr. Yaegashi, who was a senior manager late in his career and did not expect to lead a major engine development program again, was recruited by a board member to lead the hybrid engine team. He negotiated vigorously with the executives for control and for the best people (not money or position for himself). After explaining the situation to his wife, he moved into the company dormitory so that he could completely immerse himself in the enormous task at hand. Mr. Yaegashi was doing what so many Toyota engineers have done during the course of the company's long history—he put the company first. His top priority during the hybrid-engine development process was to sustain this commitment until the hybrid engine was completed. A personal life would be a distraction.

Chapter 10 described how outside suppliers become seamless extensions of Toyota. This extends to the way in which suppliers' engineers interact with the Toyota culture. They are expected to be dedicated to the company and have a strong work ethic, characterized by technical excellence and discipline. When Toyota decided to establish a joint venture with Matsushita to outsource the *Prius* battery technology, it first had to be convinced that these qualities were in place. Toyota's Electric Vehicle Division had already codeveloped with Matsushita a nickel-metal hydride battery for an electrical version of the RAV4 sport utility vehicle, so the company had a successful track record. Nevertheless, Toyota was concerned about the differences in the companies' cultures and whether this would interfere with producing a high-quality battery. This concern was alleviated when lead engineer Fujii noticed that a young Matsushita engineer was pale and haggard. When Fujii discovered that the reason was that the engineer had been up all night working, he relaxed (Itazaki, 1999). The dedication he saw made him confident that that Toyota and Matsushita could work together despite their cultural differences.

Toyota and Matsushita ultimately ironed out any cultural disparities and produced a world-class hybrid vehicle battery, but that is not the main point of the story. What really matters here is that Toyota values discipline and work ethic and requires these of everyone—inside and outside the company. All lean processes and tools described in this book are 100 percent dependent on this discipline and work ethic paradigm. The following examples show how and why.

- *Standardization and working to process. Kaizen* begins with a stable, standardized process (see Chapter 13). All Toyota engineers believe in the importance of standardization—investing in creating and improving on standards and then strictly adhering to them when they have been selected as the standards.
- *Maintaining schedules.* It is easy to use project planning software to make beautiful charts with detailed timing, but Toyota engineers actually take them as gospel and follow them precisely. For instance, it is not entirely unusual for Toyota engineers to sleep near their worksites to ensure that they start or finish something

exactly on time. Consider some of your most important deadlines, the deadlines you would move heaven and earth to meet. Toyota engineers view *all* schedules and deadlines in this way.

- *A3 disciplined communication method.* The process of boiling a project down to the essential facts and creating a visual one-page report is excruciating. Americans who work for Toyota report that this is one of the most difficult and at times frustrating processes to learn. You must have disciplined workers who have an absolute commitment to the process, no matter how uncomfortable and onerous that process is.
- *Nemawashi.* A key part of A3 report writing is the process of getting consensus while the task is in progress. Meeting after meeting with person after person seems tedious and wasteful. If this procedure is not ingrained in the culture, engineers will quickly begin to find shortcuts, but these inevitably defeat the purpose.

Following all of these detailed processes involves intensive communications and meetings and takes a lot of time and goes well beyond a 9 to 5 mentality. When Japanese coordinators come to the United States to work, they often leave their families in Japan. Then they do what they did in Japan—work long hours. This sends a strong and sometimes disturbing message to their American colleagues that, for Toyota employees, there is little life outside of work. Most Japanese businesses do not have "work-life balance" programs. But Toyota soon learned that other countries, like the United States, have their own culture and realized that it would have to adapt to this by developing a hybrid culture that would accommodate cultural differences without destroying either parent culture in the process.

In the United States, for example, family and personal interests are important—life does not begin and end with work. This does not stop Toyota from pursuing its commitment to discipline and the work ethic in its U.S. affiliates. Toyota looks for and finds engineers committed to their craft, individuals who love technical work and have a strong work ethic with a high degree of discipline. Moreover, Toyota leadership reinforces this by its commitment to creating excellent engineers.

Everyday *Kaizen*

One phrase that is constantly heard at Toyota is, "It's how we work, how we do our job every day . . . each time a bit better." The phrase is a reflection of Toyota's commitment to *kaizen*. *Kaizen is not a technique; it is a passion and way of life*. Without a culture of *kaizen*, there can be no lean product development process. But what does this mean and how can it be translated into processes? A simple way is to show what happens during a PD process. Engineers set specific, measurable, component-level goals to build continuous improvement into each program, and every program is an opportunity to improve over the last program. Every stage in the program is a learning opportunity. Everyone learns from the first prototype and does a better job on the second prototype.

One thing inherent to the *kaizen* spirit is humility, because once you believe you are the best and unbeatable, *kaizen* is dead. In 2004, Toyota reached what most Japanese and U.S. companies saw as the pinnacle of success—over one trillion yen in profit ($10 billion) and growth, year in and year out. Nonetheless, President Fujio Cho perceived a threat and declared a crisis. The threat was complacency. This was not an anomaly; it is simply one example of how Toyota constantly needs to reinvent itself (Automotive Industry Management Briefing Conference, Traverse City, August 2004):

> Why in the world would we want to re-invent ourselves when business is good? Because any company not willing to take the risk of reinventing itself is doomed. The world today is changing much too fast. If you are not busy reinventing your company, I guarantee you are falling backwards. Even worse, your customers are probably looking elsewhere.

During this same time of unparalleled success at Toyota, we found the spirit of *kaizen* not just with the president, but throughout all levels of Toyota. It was very common for engineers and managers to tell us about all the challenges they faced and how much work remains to be done in their particular area of responsibility. We do not recall a single incident where a Toyota employee, at any level, communicated a sense of complacency.

Desperate attempts to implement *kaizen* to reduce cost to save a company on the brink of bankruptcy is something any company would be willing to try, but this "survival instinct ploy" is a poor indicator of how *kaizen* works. The real test of *kaizen* is what a company does when it is on the brink of unprecedented success. At Toyota, engineers and managers alike have told the authors that "really we are not so good, we have many problems still," afterward reciting an extensive list of hard work yet to be done. This is the humble spirit of lean culture.

Customer First Spirit

At Toyota, one common denominator aligns everyone to the same goal. To determine what this common denominator is, consider the following questions:

- How can Toyota make a matrix organization work? In most companies with a matrix, engineers have conflicting allegiance to the functional boss and the program manager.
- How do different Toyota chief engineers avoid competing for the best resources?
- How does Toyota use metrics and incentives to direct engineers toward common goals?
- How do Toyota engineers work cooperatively with styling instead of fighting over appearance versus function?
- How do product engineers work cooperatively with manufacturing instead of pursuing only product engineering objectives?

The answer to all these questions is the same. It is *customers come first*. There are many examples that illustrate how important this concept is to Toyota; the one presented here should suffice to show how overarching this concept truly is. In an interview with a group of Toyota body engineers, the authors asked about their early interaction with stylists from the design studio in the *kentou* phase of any particular program. In this phase, designers responsible for styling present various clay models to engineering and engineering anticipates problems they will have from an engineering and a manufacturing perspective. They develop many study drawings to explore solutions. When asked how they resolve differences when the best engineering solution does not correspond to the most artistic design, the engineers were more than a little surprised. They responded to this "strange" question by explaining styling and engineering functions in terms of customer satisfaction. Styling, they observed, is trying to give customers an attractive car they will feel proud to own. Engineering's job is to realize Styling's vision without compromising aerodynamics, functionality, or producibility. Engineers work hard to maintain the integrity of styling—because their direct customers are the stylists.

Not long after this, one of the authors gave a presentation on Toyota's system at the design studios of one of the big 3, recounting the story described in the previous paragraph. After the talk, a senior designer approached him and said: "I almost fell over in my chair when you told that story. Engineering at our company is king. If our styling causes problems, they simply change it. They would never say we are their customer." Ironically, engineers in the same company have told the authors that their styling colleagues think they are "king" and will not yield at all to make a design manufacturable. This is clearly a cultural issue with neither party willing to give in for the sake of the ultimate customer—the person who buys the car.

Learning DNA

The learning organization was discussed at length in Chapter 11 and is revisited here from a distinct but corroborative perspective. Toyota strongly believes that the capacity to learn is the main source of competitive advantage and that continuous improvement is about learning. Toyota has two major cultural biases with regard to learning:

1. *Learning is tacit.* This is the most important one. By definition, you can only transfer tacit knowledge when there are dense ties under the guidance of a skilled mentor. At Toyota, every leader is a teacher—personally training and coaching junior people in the Toyota Way.
2. *Learn by doing which means trying.* You cannot learn by theoretically determining the best way and then executing only the best way. There are many possible solutions, and you can only learn by trying them, enjoying your success, and reflecting on your failures. If you are always trying to figure out the best theoretical solution, you will be in a constant state of waiting, missing many opportunities to learn.

Toyota leaders often refer to this learning-by-doing way of thinking as part of the Toyota DNA. It is in employees' genes to try out options and learn from actual experiences. Leaders are guides, encouraging and watching for the right opportunities to impart significant lessons. This applies to all associates and to outside partners as well. Andy Lund, an American program manager for the *Sienna* minivan at the Toyota Technical Center, explained how he learned this from a Toyota Japanese quality engineer before joining Toyota:

A TMC Japanese quality engineer came to our plant from Toyota to conduct an audit. He explained that the value of a bad part is ten times the value of a good part because you will learn so much more from that bad part on how to prevent the problem from recurring again. If we go to a line and there is no bin for bad parts—a Toyota engineer will ask: "Let me see your bad parts." In the early stages of launch of *Sienna*, we have in effect a 100 percent return policy. If there are any defective parts on the *Sienna*, we request the dealers to return the defective parts back to the plant so that the Toyota quality engineers can review the actual part. That is part of *genchi genbutsu*—looking at the actual part, not just reading the field report, which we do first of course.

Accountability and Responsibility

Teams, teams, and more teams are almost an obsession at Toyota. Credit goes to the team. Teamwork is the key to success. Paradoxically, however, there is a saying at Toyota that "whenever everyone is responsible, no one is responsible." Every engineer must be accountable—everyone must deliver. At Toyota, no one wastes time blaming or criticizing others, but in the end, someone is responsible if something does not go right and that someone stands up and takes the blame (or accepts responsibility) for failure.

This willingness to accept responsibility is the spirit of *hansei* (see Chapter 11) at work. When a program is running late because dies were late, some individual will say it is his or her responsibility to get the dies done on time. If a component part did not work out, some engineer will take responsibility for not considering alternatives carefully. *Hansei* is about reflecting, identifying things that did not go well, and then taking responsibility. As one senior executive explained: "You need to feel really bad and promise never to make this same mistake again." Feeling bad and sincerely committing to doing better in the future is the driver that sustains *kaizen*. As the same executive observed, "You can not have *kaizen* without *hansei*."

Team Integrity

In a company that views learning and improving its people as essential to create and sustain a competitive advantage, the main job of leaders is to invest in teaching. Part of this is knowing how to treat the people being taught. A corollary to this is knowing what *not* to do:

- Don't lay off people at the first sign of each business downturn.
- Don't pit people against each other so you can reward the winners and turn off the losers.
- Don't leave new employees to their own devices or ambitions to learn on their own.

The good old days of lifetime employment seems to have largely disappeared from most businesses in Japan. At Toyota, however, this tradition is alive and well. It is unlikely that anyone knows or remembers the last time a full-time Toyota employee was fired. What is known is Toyota's policy of dealing with employees who stop adding value. These people get a "seat by the window," meaning they get assigned to a job where they can do no harm. A manager, for example, to whom no one reports, has a seat by the window and is assigned tasks that cannot injure the company or anyone it. While some might see this as a "cushy job" with few headaches, in the culture of Toyota, where everyone feels self-imposed pressure to perform at a high level, this is a very embarrassing position to have.

In choosing this solution and other solutions aimed at protecting its integrity, Toyota maintains and nurtures the "towering technical competence" discussed in Chapter 9. Developing this competence is a very structured process, with a common set of socialization experiences, imprinted upon new employees as soon as they join the company. Some might call this "indoctrination" and shudder at the thought of individuality being stamped out. To an extent, they would be correct. It is indoctrination, but its objective is not to eliminate free and creative thought. The true objective is to indoctrinate each employee into the Toyota Way while supporting and encouraging the creative thought needed to approach problem solving or to offer new perspectives—within the context of how Toyota gets work done.

To achieve this, certain tenets of the Toyota Way are "grafted" into employee DNA. For example, engineers must readily learn about and accept *genchi genbutsu, nemawashi, kaizen,* the value of people and partners, the role of the leader as a teacher, and the spirit of challenge can all be considered DNA genes. If everyone has a different perspective on such core issues, there will be no Toyota Way, and Toyota will lose the "golden egg" that gives it a competitive edge.

What keeps Toyota teams aligned are common objectives, beginning at the top and cascading to the bottom. The *hoshin* planning process (discussed in Chapter 15) determines goals company-wide; every employee has objectives that are developed with his or her immediate supervisor, and these objectives dovetail with the objectives of the next level up. Pursuing individual objectives aligns individuals into teams working toward company objectives. This is possible only in a culture that accepts working toward objectives outside one's own individual interests.

Managing Upward, Downward, and Sideways: *Hourensou* Management

Unfortunately, many modern-day engineering managers believe their role in an organization is to attend meetings, keep abreast of the latest organizational politics, make the tough decisions about the big problems in the company, and generally look upward and outward. The philosophy seems to be that a good manager is good at delegating, and good engineers should work autonomously.

The principle of "*hourensou*" suggests a very different image of the Toyota manager. This Japanese management concept can be interpreted as Hou (*hou koku*—to report), Ren (*renroku*—to give updates periodically), and Sou (*sou dan*—to consult or advise). In other words, Toyota leaders have the responsibility of staying informed about the activities of subordinates so they can report on key activities, give updates to their leaders, and advise subordinates.

We saw this in action when George Yamashina was president of the Toyota Technical Center in Michigan. Yamashina seemed to be everywhere at once, wandering about, asking questions, talking to technicians. All of his managers were required to send him an e-mail at the end of each day, summarizing in bullet points the key activities for that day. He got over 40 messages a day from his subordinates, and, reading them, particularly in English, was time consuming. Nonetheless, he believed it was a very worthwhile investment and read each one.

Yamashina had a broad view of the entire organization. To see the big picture, he was constantly in motion, traveling from the Technical Center in Ann Arbor to manufacturing sites where resident engineers were located to the Arizona proving ground, to headquarters, to Japan, and back in an endless cycle. When questioned about this, Yamashina explained that by seeing the pieces, it was easier to see the whole. For example, if one engineer in one part of the organization had already had run tests with negative results, there was no reason for another engineer in a different part of the organization to run the same test and discover the same thing. The test result of the first engineer could be provided to the second engineer and that would be that.

The authors had an insightful conversation on the subject of *houren-sou* with Yamashina and several American Toyota managers. One of these was Chuck Gulash, vice president of the Toyota Technical Center, who admitted that he was at first turned off by this practice:

> Initially I was rejecting the concept because it seemed like micro-management. Now I fully support the activity and I think I under-stand it. I always believed in management by walking around, but what was it I was supposed to be learning by doing that. So what? You walk around. *Hourensou* provides an efficient and disciplined way of knowing what is going on for me as an executive and allows for sharing across the organization.

This quote vividly reflects the clash of Toyota's culture with a typical U.S. management culture. In the United States, "nosing around" is being a bus body. It is a form of micromanagement. An effective manager should delegate and then stay out of the engineer's way. For a Toyota manager, this is simply abdicating responsibility. How can you consult and advise if you do not know what is going on? If you have no more information about what is going on across the organization than any other engineer, what is your value?

The Right Process Will Yield the Right Results

Information gleaned during the authors' interviews with Toyota engineers and managers in Japan and the United States shows that, among those who are Japanese, many aspects of the lean culture are engrained and have a direct influence on how they think and act. The three lean culture subsystems—process, people, tools and technology—are truly part of their DNA. Toyota has done well in indoctrinating U.S. managers and workers in lean thinking, but getting these people to live and breathe "lean thinking"—which is what their Japanese counterparts do—is a bit of a struggle. Japanese managers believe unwaveringly that the Toyota Way is the right way. For them, there is no question that: *There is a process for doing everything and following that process will lead to positive results.* Cutting corners may work from time to time but will not consistently lead to excellent results.

Most U.S. managers do not believe with the same conviction until they see incontrovertible evidence that makes them believers. Andy Lund explained the tensions he sometimes felt as *Sienna* program manager between the Toyota Way and the expedient way. For example, part of his job was to lead the *hansei* at the end of the program. In his role as program manager, he led sessions devoted to reflecting on the program and had to gather a lot of information and use the five-why method to analyze the root causes. He met with engineer after engineer to listen to their points of view on what had happened; frequently, he wondered whether talking to yet another person would be of any value and was tempted to skip some people. But each time he talked to a person he had considered skipping over, he learned something surprising and informative. Lund's struggle was cultural: a battle between Toyota's belief in the process and his American roots, which urged him to reach a conclusion quickly. What emphasizes the depth of this struggle is the fact that Lund grew up in Japan.

The Culture Supports the Process

Many automakers would like to adopt Toyota's lean PD process. Some have instituted major initiatives to learn from Toyota and are all too often disappointed. They have tried A3 reports, functional build (screw body build), a CE role, *hoshin* planning, and a know-how database. Each of these tools has been helpful for a while, but their effect has eventually dissipated and the tools themselves have fallen into the black hole of disuse. In most cases, engineers and managers lost interest or returned to old ways that more or less work. The main reason companies have so much difficulty adopting and sustaining these simple, common sense principles is culture. Figure 12-1 compares the cultures of Toyota and NAC, the company that has been used throughout this work as the antithesis of true lean thinking.

A quick glance shows how diametrically opposed Toyota and NAC are on every critical dimension. NAC is a business driven company; meeting

Toyota	NAC
Technical excellence	Business excellence
Process discipline and work ethic	Results focus
Kaizen every day	New initiatives
Planning and detailed execution	Just do it
Learning DNA	No problem

Figure 12-1. Contrasting Cultures at Toyota and NAC

expectations for quarterly profits for Wall Street is the highest objective, so it has a finance-dominated culture. Technical excellence is secondary because NAC is about getting bottom line results in any way at any cost. Where Toyota reveres technical excellence and invests heavily in the development of its people, many NAC executives are fond of saying that "costs walk in on two legs" and focus efforts on people reduction, even if this may lead to losing core competencies. Toyota is about the process. Follow the right process and you will get the desired results. NAC is often motivated by the latest, cutting-edge initiatives and technologies that can make work easier and faster—shortening the line between two points. Forget about the daily mundane activities, like detailed engineering, or building cars that are not exciting. In contrast, Toyota is about continuously improving mundane processes. Engaging in detailed planning to an almost compulsive degree is the norm at Toyota. NAC's result orientation leads to a "just do it" mentality, without an effort to capture any learning to leverage for the next program. Toyota is all about learning, program to program, engineer to engineer. At NAC, asking for detailed explanations of what has been done before is considered rude and a sign of incompetence. A quick smile and a reassuring "no problem" is typical of NAC engineers, no matter how complex a task is assigned. The same behavior at Toyota would raise eyebrows. How can you do a good job in the Toyota Way without deep questioning, *nemawashi*, going to see for yourself, sharing insights with the team, and learning as an organization?

Given the culture at NAC (and similar companies), it is no real wonder that implementing and sustaining Toyota-like development processes can be nearly impossible. Three examples of what tends to happen are presented below.

Chief Engineering System. In the LPDS Principle 5, you *develop a chief engineer system to integrate development from start to finish.* That is, the chief engineer drives the entire product development system—this is a foundation of the lean PD system. Several traditional auto companies recognized this and ordered that their company create chief engineers. In a bureaucratic culture, this is easier said than done. The typical response was to change the job titles of program managers to chief engineers and give them some training to explain their new roles so then they could become official "chief engineers." This, of course, raises at least three questions: Will the new CEs have the technical expertise and organizational respect to get the job done? Will functional bosses assign the right people to programs at the right time? Will people who are evaluated by their functional bosses and also report to a chief engineer on a program follow the instructions of one or the other? It is not difficult to see why this process is untenable. It is not surprising that the failure rate of these "instant CEs" was quite high at all of these companies.

Front-load. LPDS Principle 2 states *front-load the product development process to thoroughly explore alternative solutions while there is maximum design space.* By front-loading, you anticipate and solve problems before tools are even designed and before any capital investment. Toyota uses a *kentou* phase to do this. In the result-oriented culture of NAC, new questions arise: Will the *kentou* phase be taken seriously? Will engineering managers assign the best engineers to work on the *kentou* phase? Will engineers have the time, know-how, or discipline to work through each and every drawing, root out possible problems, and put in countermeasures? Will they overcome functional boundaries and turf mentality to work collaboratively across functional organizations? In a "just do it" hurry-up culture, will broad circulation of the *kentou* lead to deep reflection on each detailed design feature? The global answer is "not likely."

Level the workload. In the LPDS Principle 3, *create a leveled product development process flow*, leveling the workload depends upon flexible staffing as the program proceeds. At peak points in the program, you need to pick engineers and technicians from flexible pools and from suppliers. Here again, NAC's culture elicits some interesting questions: Can the company flexibly recruit engineers who will seamlessly contribute to the engineering process? Will this pool of engineers be properly trained in the "NAC Way" to understand the design process and design philosophy? Will engineers from this pool speak the same language as full-time career employees? Will they be as motivated as NAC engineers to contribute to the common goal of putting the customer first? Again, not likely.

As this brief analysis illustrates, no amount of top-down mandated process changes or lean tools will penetrate the wrong culture and create the conditions needed to make lean tools and processes effective. Plopping the process and tools into an existing, alien culture guarantees rejection, in much the same way that a body will reject an organ transplant when the organ is incompatible with the body's other organs or systems.

Leaders Renew the Culture

Companies seriously interested in emulating Toyota's lean PD design often ask the following question: "Once we have put these lean tools in place how can we sustain them?" Implicit in this question is the understanding that adopting a system does not necessarily mean the system will work and continue to work over time. There is no guarantee that once a company puts the effort into lean, the tools will take on a life of their own and sustain themselves. Whether these tools thrive or wither often depends on leadership. Moreover, it is important to remember that leaders get the culture they live and tolerate.

Within the Toyota system, leaders serve as the bearers of the Toyota culture and exemplify the culture in their daily behavior. It can certainly be tedious to follow the right process day in and day out. No Toyota leader has ever assumed that lean sustains itself. On the contrary, the company's leaders intuitively understand that the natural evolution of a culture will atrophy and deteriorate unless its leaders are continually renewing and passing the DNA to others. Expressed in terms of thermodynamics, systems tend toward the lowest state of energy. Only by adding new energy to arrest entropy (atrophy), can you maintain or improve a system. Leaders are the primary source of energy; they arrest the atrophy of lean tools and keep them thriving and evolving in the culture. It is they who can sustain lean thinking.

This chapter concludes the five LPDS principles for the people subsystem. The next section of this work examines the third subsystem, tools and technology, and the final principles, 11 to 13.

LPDS Basics for Principle Ten

Build a culture to support excellence and relentless improvement

Lean tools will not be effective unless used in a supportive culture. Culture is the way work is done and the way people think about the work and products. Some of the core cultural values that support lean product development are *genchi genbutsu*, set-based thinking, *hansei*, and putting the customer first. This broadly-shared cultural DNA is fundamental to the success of lean thinking and a further reason why it is a challenge, even within Toyota, to teach the lean product development system to new employees globally. Key characteristics of Toyota's high performance culture, which should be used within your own organization, include the following:

- Technical and engineering excellence must be highly valued.
- The culture must be based on discipline and a strong work ethic.
- Improving through *kaizen* every day must be engrained as the way to do work.
- Everyone involved in the development process must have a customer-first spirit.
- Learning as an organization must be engrained in the company's DNA.
- Individuals must be willing to stand up and take responsibility when things do not go well.
- Investing in engineers and treating them like valued assets must be the norm.
- All engineers must step up to challenges as a matter of course.
- Strictly following the right process for doing the work must be highly valued.
- Mistakes must be viewed as learning opportunities.
- Leaders must be the culture bearers and lead by example every day.

Tools and Technology Subsystem

CHAPTER 13

Adapt Technology to Fit Your People and Processes

The first rule of any technology used in a business is that automation applied to an efficient operation will magnify the efficiency. The second is that automation applied to an inefficient operation will magnify the inefficiency.

BILL GATES, president and CEO, Microsoft

COMPANIES AROUND THE WORLD are trying to find ways to accelerate product development, seeing this as a way to improve competitive advantage. In many cases, their efforts to speed up the PD process focus on advanced technology. But using rapid prototyping, digital simulations, product life-cycle management, virtual engineering, and similar tools and technology to revolutionize PD may not yield the hoped for result, primarily because technologies are seldom exclusive. Any company can copy or purchase the tools and technology used by any other company. Successful utilization of such tools and technology depends on the ability to customize them in a way that makes them exclusive and integrates them uniquely to the company using them. No one, for example, can deny that that tools and technologies have had a significant influence on Toyota's ability to achieve development cycles of 15 months and less. But it is important to recognize that this occurred because Toyota has had the foresight and discipline to customize tools and technology to fit within a broader framework, one that includes people and processes.

Five Primary Principles for Choosing Tools and Technology

Finding one's way through the "technojungle" is no easy task. Rapidly changing functionality along with the jungle's many hidden hazards makes identifying and choosing the correct path difficult. Decisions about which tools or technology to adopt and when and how to integrate them into the organization have significant implications for a company's PD system. The selection process requires a substantial investment of both financial and human resources and can result in organizational upheaval and the loss of irreplaceable time, especially if the new tool or technology does not integrate well with the other two LPDS subsystems, process and people. LPDS Principle 11, *adapt tools and technology to fit the people and processes*, is a prime directive for implementing and sustaining a successful lean PD system, one that Toyota has internalized exceptionally well. We subdivide this principle into five highly effective steps or subprinciples.

1. *Technologies must be seamlessly integrated.* Toyota integrates many of its product development technologies into its V-Comm system. V-Comm integrates surfacing/design testing, digital assembly, simulation and the know-how database into a single seamless system, which enables Toyota engineers to move from design to ergonomic simulation, to update testing results provided by colleagues, and to have access to necessary standards and checklists. This powerful suite of tools and real-time information is made available to the engineers working on a program.

2. *Technologies should support the process, not drive it.* Technology consultants often advise, "You need to constantly change your company's processes to keep up with the latest technology." From Toyota's perspective, this is backwards. Changing the process to conform to technology leads to instability, drives massive process variation, confuses people, and creates tremendous waste. As companies tinker with processes to force maximum results from some new whiz-bang technology to demonstrate the wisdom of their strategy and investment, they waste time and money. All too often, the state-of-the-art technology becomes obsolete within a single year. What ensues is a mad rush to acquire the next technological fad, a move that is often even more detrimental to an already dysfunctional process.

3. *Technologies should enhance people, not replace them.* In many companies, the primary justification for major technology purchases is reduction of labor costs—that is, how many people will this replace? In a business that is talent driven and dependent on technical expertise, this is clearly counterproductive. In product development, it is best to choose tools and technology that make the best use of engineers' time and talents. Tools and technology should never be seen as substitutes for expertise; they should complement expertise.

4. *Specific solution oriented: not a silver bullet.* Technology can provide high-leverage if a company has a clearly defined purpose for it. Searching for a nonexistent Holy Grail is futile. Technology is never a substitute for the hard work that is required to make a product development system competitive. Its potential lies in supporting and accelerating that hard work once a lean process is in place and highly skilled people are appropriately trained and organized.

5. *Right size—not king sized.* Many western companies have a tendency to buy the biggest, baddest, fastest, and newest tools on the market. Our old friend NAC, for instance, often boasts that it is going to leapfrog Toyota by technological one-upmanship. This, however, seldom happens and the following example illustrates why. For decades, Toyota has successfully used notebooks for its engineering checklists. NAC developed an impressive online and fully integrated database, convinced that this innovation would help it zoom past Toyota in knowledge management. But the data was vacuous, owned by an independent "technology group," and rarely used. The point is that electronically storing data is no substitute for engineers who develop knowledge over time, create checklists to capture the knowledge, and rely on the cumulative knowledge and the checklists to ensure that things are being done properly.

Technology in Lean Product Development

Successful automotive product development relies on the successful application of hundreds of diverse technologies. A detailed analysis of those technologies is far beyond the scope of this book and this section covers only a selected few that are central to the product development process. The purpose here is to show how Toyota adheres to the principles discussed above, avoiding technology pitfalls. To accomplish this purpose, the sections below are structured as a series of contrasting models (Toyota and NAC) and how these mesh or fail to mesh with lean principles. It is important to introduce this discussion by emphasizing that a lean environment enables Toyota to 1) find the tool that fits, 2) make it work seamlessly in already effective processes, and 3) align the technology and tools to enhance both product development and manufacturing.

Digital Engineering at Toyota

Toyota has leveraged an integrated suite of digital tools all along the product development value stream. They began digital engineering implementation in Product Engineering and now have developed powerful tools to enable engineers everywhere in product development. This integrated suite of tools, commonly referred to as V-Comm, includes their design and surfacing software, digital assembly software, numerous databases including the Know-How database and database of standard processes, the assembly sequence, as well as a number of communication tools.

Toyota has been progressively implementing digital engineering for the past decade. Beginning with design tools for instrument panels, stampings and vehicle body assembly in 1996, they progressed through engine compartment and added supplier-designed parts by 1998. In 1999, the Scion bB or black box program was the first vehicle to employ a full suite of digital tools.

More recently, Toyota has adapted a new commercial design software (CATIA V-5) and has also made dramatic enhancements to their Digital Assembly software, allowing them to do even more work very early in the development process.

Design Technology at Toyota

Toyota spends a significant amount of time customizing the design software that it uses, ensuring that this software fits its process and methods before it is used in a product development program. For instance, at Toyota, the ability to design a part within its working environment and to view as much or as little of that environment as needed for a complex product is critical to software design. As noted in Chapter 5, one of the key elements of Toyota's design process is to resolve integration issues and achieve system compatibility before completing designs. In this light the benefit of this technology is obvious. Designing in an integrated product environment allows Toyota engineers to view as much of the surrounding design and as many design variations as they wish. With a single command, such as "show all parts within some distance of my part," the software lets an engineer see what he or she needs to see within the design environment. Furthermore, by selecting specific feature codes from the bill of material, engineers can view variations based on option packages and the like. This helps reduce late engineering changes. Moreover, as part of the seamless integration, and improved communication the factory and dealership can use the same feature codes to maintain accurate communication throughout the enterprise.

Toyota's design software also supports the LPDS Principle 4 "of rigorous standardization to reduce variation." By utilizing associative CAD models of parts, Toyota engineers can easily morph standard construction sections to fit new design parameters based on preset design rules. For example, a roof bow (roof reinforcement) section (edge view) is predetermined based on analytical modeling for roll over, etc. As new roofs take a new sweep to fit the specific styling requirements of a new vehicle, engineers can morph that roof bow to fit the new roof sweep while maintaining critical section proportions.

Engineers can also apply this feature to die and fixture designs, morphing standard components of the tools to fit new part design requirements. This saves time in both product design and tool design. At the same time, it increases performance reliability, further supporting Toyota's core principle of reusability and allowing for better integration of design and manufacturing.

A final important characteristic of Toyota's design software is that it is parametric. That is, as changes are made by one engineer that affect another engineer's design (whether product or manufacturing), related parts are adapted accordingly. The changes are highlighted in a different color and assigned a change code, alerting the second engineer that a change affecting his or her design has been made. This engineer can then review the data and contact the first engineer immediately if something seems amiss. In practice, this clearly supports LPDS Principle 3 because leveled flow depends on seamless cross-functional execution that prevents a lot of rework loops. It also supports an engineer's ability to design in context "real time" and synchronizes concurrent engineering.

Virtual Manufacturing and Digital Visualization at NAC

We can appreciate the strength of Toyota's approach to integrating technology, processes and people by contrasting it to NAC's use of virtual manufacturing and digital visualization. Designed to utilize new design and surrogate design data to check the effects of tolerance stacking for part clearances/interferences and manufacturing access for assembly, it can also be used to study manufacturing cycle times.

From a vehicle assembly perspective, NAC employs this technology after design data is released as a way to check design accuracy and completeness. Engineers schedule virtual build events in a large room where representatives from various disciplines gather for several days to review issues. This culminates in a summary review by senior leadership. Unfortunately, the virtual build events are scheduled late in the design development process and are completely controlled by an isolated specialist group with no other deliverables. As a result, engineers view the virtual factory technology as just one more auditing tool that identifies problems. Lots of issues are reported, but because they are reported late in the process, it is often too late to do anything more than compromise on design changes. Rather than learning from the tool and using it to benefit future designs, engineers see it as a tool that leaves them vulnerable to criticism.

Toyota uses this technology at the front end of a process to identify problems early. But it is important to note that it is not the software that identifies problems but the engineers. The software is merely a tool that generates virtual representations that engineers with deep knowledge can use to identify problems and solve them. At Toyota these virtual tools, combined with towering technical competence and excellent problem solving processes, enable LPDS principle two: *Front-load the PD Process to Explore Alternatives Thoroughly.*

Digital Assembly at Toyota

Since the early 1990s, Toyota has employed *Digital Assembly* (DA) during the very early *kentou* phase of development and beyond. By utilizing early scan, standardized design data, common components, and actual 3-D manufacturing work station data, Toyota studies part clearance/body fit up, ergonomics, assembly, cycle times, work station design, fixture net surfaces and locators, as well as craftsmanship (interior and exterior vehicle fit and finish) issues. At Toyota, the work cells, stations, presses, etc., are actual 3-D geometry from the factory floor, showing precise clearances and constraints that workers will face on the factory floor. Among other things, Toyota's Digital Assembly allows engineers to:

- Study how individual components will be assembled into a vehicle and identify potential design interferences long before designs are completed. They can also see the effects of various tolerance-stacking scenarios and design to maximize best-fit conditions and design in high levels of craftsmanship with specially developed *Digital Assembly* (DA) software.

- Use *workability* DA to study in detail the effects that design changes will have on ergonomic issues involved in assembling a vehicle. By utilizing DA in conjunction with the assembly plant pilot team (hourly workers assigned for two years to work on preparing for new vehicle launch) during the *Kentou* period, they can address both current as well as anticipated human factor issues and identify the ergonomically safest and most efficient way to assemble the vehicle.
- Study detailed processes, such as welding or stamping, utilizing data from a new product. Manufacturing engineering can determine welder access, cycle times, and even assembly work station layout in parallel with product engineering.
- Optimize fixture designs using the process planner function of DA by identifying potential part deflection points and access issues.

In the past, Toyota was unable to study many of these issues until the first physical prototypes, which, in its view was far too late in the process because changes were expensive and disruptive. DA resolves both of these issues and enables designed-in countermeasures before the program hits prototype. This generally results in fewer prototypes (in some cases none) and better quality product saving time and money. This, in turn, allows Toyota to not only start earlier but to exactly match the new product design requirements to each specific plant in which it will be manufactured.

Further, by putting the power of this technology in the hands of the individual engineer, Toyota is also adhering to its people principles. The technology is readily available to design and manufacturing engineers alike as a part of how they do their work. The availability of these data also enables engineers to utilize actual design and real world manufacturing operations information to begin basic Industrial Engineering tasks very early in the process. Engineers can check sequencing, operating time estimates, stack operations to check against takt times and conduct virtual simulations to help make design decisions that optimize manufacturing operations. The result is a technology that helps to drive lean manufacturing to the very beginning of the development value stream—enabling a *lean from the start mentality*.

You may recall that in Chapter 6 we discussed how the power of standardization enabled Toyota to build eight different body types on the same line without having to build, maintain, and change expensive pallets for each body style. These digital tools are critical to maintaining that capability on new vehicle designs. In addition, by sharing digital information throughout the enterprise, Toyota can bring multiple functional experts as well as production associates together virtually to work with specific part design data, detailed process information, actual manufacturing environmental factors, as well as detailed quality and performance histories to improve processes and optimize designs for all facets of manufacturing, including the body shop, final assembly, paint, material handling and supplier shipments.

By employing the technology early in, and consistently throughout, the process, Toyota is also adhering to LPDS process Principle 2 on frontloading. It is also making the technology seamless with other technologies that engineers use. DA is also the primary technology employed at design reviews. So, in addition to engineer-to-engineer communication, DA supports full-team or subteam integration.

Finite Element Analysis at NAC and Toyota

Formability finite element analysis (FEA) is a crucial tool for testing structural properties. In the world of designing dies to stamp out body parts, companies use FEA to predict the formability of the metal and the effectiveness of the draw die. The draw die is the first (except for the blanking die) and primary metal forming die in the manufacturing process and has a significant impact on the rest of the manufacturing process. Consequently, it is important to get this die right. Formability FEA software can predict material stress, strain, compression, thinning, and fracture during the forming operation.

An animation function allows users to see when a defect is occurring in the forming process and also shows material feed during the forming process, which can be replicated during the actual die tryout process. Based on the results of the formability FEA, engineers can modify part or draw die design to optimize part forming before dies are made.

As powerful as formability FEA can be, its potential has not been fully exploited at NAC. Despite the 100 percent application of formability FEA, there is still a great deal of formability failure at die tryout. One reason for this is that NAC does not standardize inputs to the FEA programs. In addition, there has been relatively little effort to develop statistically significant correlation studies between FEA and actual die tryout. To its credit, to make the FEA tool more effective, NAC has worked at addressing both these problems. Because NAC does not have strong process or design standardization, the company's stamping engineers use FEA to test unique part designs and processes, and the tool does help in this capacity. Moreover, anything that makes FEA more effective will undoubtedly have a significant positive impact on NAC's manufacturing engineering process.

Toyota also uses formability FEA, but on a much more limited basis. Fewer than one-third of Toyota parts go through the FEA process. Toyota uses the FEA tool primarily for exceptional or unique parts—an extra step that it takes only if absolutely required because running FEA takes time. Because it scrupulously and rigorously implements process standardization and high-level design, Toyota seldom needs to predict the formability of most parts. Applying FEA to standard parts would be a form of waste. Even so, Toyota has spent a good amount of time developing input standards for FEA use, and detailed checklists guide both FEA setup and results interpretation.

Tools for Manufacturing Engineering and Tool Making

People often think of the development process in terms of its eventual impact on manufacturing. In reality, there is a large amount of manufacturing in any product development process; the integration of product and process design can make the difference between fast cycles with processes that work right the first time and long cycles with a great deal of engineering rework at launch. In fact, as Clark and Fujimoto (1991)[1] have suggested, strong manufacturing and manufacturing engineering skills are critical enablers of excellence in product development. How tools and technologies for production preparation are used in early product development matters tremendously. The following contrastive study of how tools and technology are used in manufacturing engineering and in tool machining, construction, and tryout at Toyota and NAC illustrates why.

Checklists and Standardization Tools at Toyota and NAC

Throughout this book, the importance of rigorously applied engineering checklists (both for product engineering and production engineering) has

---

1. Kim B. Clark and Takahiro Fujimoto, *Product Development Performance: Strategy, Organization, and Management in the World Auto Industry*, Boston: Harvard Business School Press, 1991.

been emphasized. At Toyota, this also applies to manufacturing engineering and tool making. In fact, specific process and part-based checklists are ubiquitous in the Toyota PD system. By and large, Toyota incorporates these checklists into the macro know-how database, which functional specialists update at each program's production preparation stage. As noted above, this even applies to tools and specific software applications (FEA), and it goes almost without saying that the use of checklists extends to all aspects of die design, preparation, and tryout. In contrast, NAC has only recently begun utilizing standardization tools and checklists (attempting to learn from Toyota), and only certain individuals on special pilot programs use them rigorously. If NAC hopes to develop a lean PD system, these tools will need to become broadly owned and rigorously used.

Solids Die Design: NAC Versus Toyota

NAC recognizes the importance of solids die design and has the technical computer capability to build solid models. However, it has only recently begun to utilize it on all die designs. NAC began the solids-based die design initiative around 2001, and implementation was slow. Suppliers do all of NAC's die designs and their resistance to making significant investments in this technology has inhibited NAC's transition to solids-die design. Another issue has been NAC's insistence on working with in-house design software and developing its own ancillary applications rather than using commercially available die design software.

Toyota has been utilizing solids-based die design on all of its die designs since the late 1990s. Toyota was very aggressive in adopting this technology and recognizing its potential efficiency impact both on die design and downstream processes. In its customary way, Toyota worked intimately with the software company to customize the packages to Toyota's specifications, a beneficial relationship that ultimately also improved the software company's services to other customers. Toyota already had a well-developed in-house die design and manufacturing capability, so it did not have to convince outside die designers to use the technology.

The benefits of solids die technology are obvious. When combined with Toyota's standardized manufacturing processes, solids die design becomes a powerful tool for synchronizing cross-functional processes and achieving concurrent execution without rework. Because they have access to a standardized component library, die designers can select the right component "off the shelf" even before the product data has been fully stabilized. Solids die design also works with design data in its native form without the need for conversion and the associated loss of detail.

Using simulation technology supported by solids die design gives the design group the ability to check clearances, automation and fundamental die functionality. Engineers can simulate production forming, part transfer, scrap shed, and can also perform casting stress analysis. When used in combination with Toyota's checklists and standards, simulation eliminates the need for many of the traditional reviews and much of the rework common at NAC. Toyota can now design its dies with great detail and accuracy. In addition to simulating actual die performance, Toyota also utilizes simulation technology to enhance the manufacture of their dies. It utilizes this capability to simulate die machining, create detailed bills of materials for dies, and drive standardized die assembly sequences. This, in turn, enables lean die manufacturing, discussed later in this chapter. Another important advantage to solids die design is that it enables high-speed pattern building, which is an important part of the die making process at Toyota.

Pattern Making at NAC Versus High-speed Pattern Making at Toyota

Patterns are representative die shapes made of foam and used for casting dies. Pattern making at NAC has generally been a traditional craft-based process. About 60 to 70 percent of the pattern is machined; the remainder of the pattern shape (depending upon the die/part type) must be built by hand to pattern prints generated by the die design group. Pattern surface finish is generally too rough for foundry purposes and must also be hand finished.

One of the benefits Toyota derives from solids die design is its ability to support a high speed, automated pattern-making process. Toyota uses a solids die design software suite called CADCEUS to develop a pattern layering strategy that allows it to machine about 95 percent of its foam die patterns. Three or four long rectangular foam boards are cut to size and machined into intricate shapes, which are then glued together in layers to create a single die pattern. This automated process requires less than one-fourth of the time and none of the highly paid skilled craftsmen or pattern prints necessary for the conventional pattern building practiced at NAC.

Die Machining: NAC Versus Toyota

NAC has made a significant investment in several very large, state-of-the-art Computer Numerically Controlled (CNC) machines. These machines are capable of machining dies up to 4,000 mm × 2,500 mm and machining at a linear speed of 250 inches per minute (IPM). Because of the increased machine speeds, NAC has reduced its stepovers to 0.5 mm, thus decreasing the amount of hand finishing required for dies without exposed (class one) surfaces.

NAC, however, has not made any other significant changes to its machining technology or to the quality of the dies machined. Within lean manufacturing, decreasing the cycle time of some processes while others have longer cycle time, means that the longer-cycle time processes will be bottlenecks. For this reason, Numerically Controlled (NC) technology has not significantly increased throughput or reduced lead time. Moreover, NAC has far fewer CNC machines (about a dozen) than Toyota has, and all of these are very large and produce large quantities of dies, sometimes significantly more than are required (i.e., over production). Finally, machining makes up only about 60 percent of NAC die construction, leaving a great deal of craft-based hand work that varies significantly among individual die makers.

Toyota has also made some very significant investments in CNC machine capability, but has purchased many machines of various sizes, primarily because dies for most of the parts on a car do not require large machines. Toyota's lead tool and die facility has more than twice the number of CNC machines and is organized in flow lines by die classification (A-E). As one would expect from a lean manufacturing enterprise, Toyota has identified part families and developed dedicated flow lines—machine lines specialized for different sized dies leading to major reductions in lead time and increases in overall throughput. Dies are scheduled onto the smallest machine possible.

The high-speed capability of most of Toyota's mills allows the company to decrease stepovers to less than 0.2 mm, and even class-one surfaces require little, if any, hand work (even on radii). In addition, Toyota has developed and patented many custom machining and cutting tools to enable significant improvements in guide pin and wear pad accuracy. Standardized, precision machining makes up more than 85 percent of Toyota's total die construction process, minimizing the amount of craft-based handwork.

Toyota also focuses on precise machine scheduling. Every detail of every die is scheduled for machining by the hour. Visual machine schedule boards (located next to the machines) are used. Actual times recorded by the operator are then reviewed each day in plant walk arounds. Required cutters, work pieces, and cutter paths arrive at the machines JIT, contributing to improved machine uptime.

Toyota's high-precision dies enable rapid die construction, minimize in-press adjustment, and consequently die tryout time. High-precision dies are also critical to consistency in die performance. Between-die setup variance is much higher than within-die setup variance (Hammett, et. al, 1999), and, according to Toyota, low tolerance (high-precision) machining of dies makes dies much more repeatable between press set ups. This facilitates fast die setup in stamping operations and is central to any lean PD system.

Toyota dies are assembled in work cells. Precision machining, standardized die construction, and lean manufacturing principles combine to create an efficient construction process that is centered around specific-purpose construction cells housed within A through E category construction bays. Cells, although not always U-shaped, contain all equipment and materials required for performing specific standardized work within equalized task times. Required materials and machined components are delivered JIT to the appropriate construction cell. The assembly steps have been dissected in great detail, and standardized work has been used to reduce dramatically the time required to assemble dies. Checklists kept in the cells facilitate standardized work and at-the-source inspection. Visual boards show how the die assembly is progressing relative to a takt time, and deviations flag an andon to call for support to get back on schedule.

Tryout Presses: NAC Versus Toyota

Because most of NAC's dies require relatively long periods of tryout time, most presses at NAC are large, high tonnage stamping presses equipped with rolling bolsters to facilitate grinding outside of the press and quick resetting of the dies. Toyota requires far less die tryout time, and most of its presses are "spotting" presses. Less expensive than stamping presses, they are used only to check basic die functionality. Toyota's press layout facilitates die transfer from spotting press to stamping press where a forty-part run-off is stamped out. Dies are removed from the spotting press via rolling bolsters onto a large die transfer mechanism and rotated to a stamping press. Any additional handling is facilitated by overhead crane. This is the basis for single minute exchange of dies to effectively use the tryout press.

No-Adjust Build at NAC Versus Functional Build at Toyota

One of the key tools utilized for both identifying and resolving individual part, fixture, and assembly issues on the vehicle body is the functional or screw body build. This is a systems approach to a part-by-part analysis of the physical parts and subassemblies that enables understanding and facilitates adjustments to the fit and finish of the vehicle body. In the physical version of this process, parts were literally screwed together into what is called the screw body (later it was riveted together). The screw body was then evaluated by teams. More recently Toyota has advanced from a physical screw body to a virtual (digital) build up of the body from "as manufactured" parts data for review. In either case, the philosophy is the same: a systems approach to delivering customer perceived value, which guides decision making within a complex environment (Hammett et al, 1999; Ward et al, 1995b). To facilitate early identification of major fit issues in the prototype phase, engineers use a preliminary functional build utilizing prototype parts. The functional build is an effective decision-making and learning aid in the body development process. However, engineers need a great deal of experienced-based judgment to execute this complex process, and effective mentoring is central to developing the right skill set.

When this process is described to engineers outside of Toyota, they often ask for the specific technical basis for deciding when product dimensions that do not match the nominal dimensions in the original database are acceptable. When told how much discretion Toyota engineers are given to make these judgment calls, they are aghast. As Chapter 9 showed, however, Toyota engineers have earned this "power of discretion" because of their rigorous training and experience.

One constant to remember about Toyota is that it is an organization that constantly changes. For example, the functional build process was a key part of Toyota's methodology for body engineering. The company recognized that stamping is more of an art than a science and that it is not possible to get the stamped parts to match precisely the nominal dimensions of the original CAD database. Achieving the nominal meant grinding dies, which is expensive and time consuming. At present, Toyota is changing its philosophy and working to achieve nominal dimensions that will eliminate the need for grinding. Toyota engineers are confident that with new simulation and modeling technologies they can get much closer to nominal than was possible in years past when they originally developed functional build. And with globalization and tools going every which way for manufacturing and the sharing of platforms, architecture, and component parts across vehicle programs, it has become more important to have the tools and parts match the database. From the perspective of competitive advantage, companies now in the process of learning functional build (and there are many) may well have to compete with a Toyota that has moved on to something substantially better.

Although NAC has experimented with functional build, its prevailing vehicle build philosophy is that dimensionally correct components (using nominal dimensions) will assemble into a dimensionally correct vehicle body. Consequently, NAC places emphasis on stamping parts that adhere to generic dimensional tolerance bands, tolerances developed for rigid body parts such as machined blocks. Product engineers and assembly engineers at NAC vigorously enforce the dimensions and tolerances, sometimes ignoring the realities of the die development process. The reality, of course, is that sheet metal stampings do not conform to tolerances developed for rigid body, machined parts. Obviously, this policy of enforcing the unenforceable leads to high costs of reworking dies, much longer die tryout times, longer development lead times, and problems at launch.

Toyota's functional-build process zeroes in on dimensional characteristics that are meaningful to vehicle fit and finish, that is, dimensional characteristics that impact the build or performance of a vehicle. Many of the dimensional tolerances required of individual parts are meaningless once the part is assembled into a vehicle. One example of this is a thin, structurally weak part assembled to a thick structurally strong part. Once the thin part is clamped to the thick part, it will conform to the thick part. Spending a large amount of time in die tryout in order to make a dimensionally correct "thin part" is wasted effort—the thin part will conform to the shape of the thicker part at assembly. Moreover, dimensional tryout represents a significant amount of money in a typical program. Over 20 percent of die costs is invested in reworking dies, which can exceed $20 million. Much of that cost is geared to achieve nominal dimensions and 10 to 20 percent of component dimensional issues account for 80 to 90 percent of those rework costs (Hammett, Wahl, and Baron 1999).

Functional build can improve final product quality while simultaneously loosening tolerances and saving time and money (Hammett, Wahl, and Baron (1999). By focusing on vehicle system quality, functional build takes a system optimization perspective using original design specifications as goals. If Toyota is having difficulty meeting a particular specification during home line tryout, it may solve the problem in a downstream operation or even change another related mating part that can be changed more expediently. Indeed, as noted, Toyota has learned so effectively from the functional build process and improved die design, surface modeling, springback prediction, as well as the precision of die manufacturing, that it is gradually moving away from functional build all together and building to true nominal dimensions. NAC has not developed this capability and therefore may still benefit by seriously utilizing functional build, which will require a cultural change among engineers.

Three-dimensional noncontact measuring

Optical scanning is one of the technological advances that has enabled Toyota to move to a virtual "as manufactured" functional build process. Toyota utilizes this technology to support functional build by scanning parts manufactured at multiple locations and then assembling them virtually in a central location even before parts are shipped to the build location for physical assembly. Engineers also use this technology to measure parts in process between manufacturing operations to determine the source of variability in a manufacturing process quickly. Toyota employs this technology in both development and production processes.

Fixtures are utilized throughout the body development process to locate parts for functioning, measurement, and assembly. These fixtures can be extremely expensive, with a single large checking fixture at NAC costing as much as $250,000. NAC has focused a tremendous amount of energy and resources on building elaborate checking fixtures. Though this can provide reams of checking data, it keeps quality auditing groups busy assessing endless measurements.

Toyota's largest checking fixtures cost less than one-fifth of NAC's and produce more usable data. Instead of engaging in part inspection, Toyota focuses more on process control and source monitoring, employing three-dimensional noncontact measuring technology that produces relative measurement of complete surfaces rather than the elaborate fixturing or independent point measurements favored by NAC quality auditors. Fundamentally, the three-dimensional noncontact measuring technology works by taking a point dense, three-dimensional "picture" of the part, transferring it into the CAD environment, and comparing the part produced to the actual design geometry. Engineers or operators can immediately see any deviation between the two. Providing complete surface measurement is more useful in process diagnostics than are isolated points, and noncontact measurement systems have been shown to be faster, shop friendly, and more accurate than point-based or laser-dependent techniques (Hammett, Frescoln and Garcia-Guzman, 2003).

Adopting Technology to Enable Process

The examples presented above show that NAC and Toyota think about technology in fundamentally different ways, with NAC focusing on the technology itself and Toyota focusing, first and foremost, on the process. NAC looks at the technology and asks, "What can this technology add to our process? Can we justify purchasing it? How can we get the most for the least amount of money from this technology?" Once purchased, the technology often requires that the current process be changed to make the best use of it. On surface, this may seem logical, but in a lean PD system, it is not. In a true lean PD system, the first step is to reduce waste in a process, then look for opportunities in which the use of technology can support the PD process and develop clear requirements the technology can meet. Finally, if necessary, you work with software vendors to customize the technology appropriately to fit the process. *In a lean PD process, rather than having technology drive changes in the process, the process leads you to adopting technology that can enhance it.* A lean organization understands that its first concern is always to work at perfecting the process; technology is useful only as an enabler or catalyst.

In manufacturing situations (e.g., making dies), NAC's approach is to invest in a small number of expensive and very large machining centers, which creates batch building, poor synchronization across steps, and bottleneck operations. Toyota, on the other hand, uses the lean concept of one-piece flow, separating parts into part families and creating flow lines. Through this approach, Toyota can synchronize steps in the process, increase throughput, and reduce lead time.

Whether it looks at solids, FEA, or functional build, NAC sees technology as a silver bullet. Ironically, this idealization promotes ineffective use of technology. Often, engineers are then blamed for not using it correctly. LDPS Principle 11, *adapt tools and technology to fit your people and processes*, suggests that a more successful approach is to have a realistic and pragmatic perspective, one that sees technology as an enhancement tool rather than as a silver bullet. It is from this position that lean enterprises adapt more modest, less costly, and more effective technology.

LPDS Basics for Principle Eleven

Adapt technology to fit your people and processes

The following guidelines can help you make advances in technologies that act as powerful PD system accelerators:

1. Integrate new technology seamlessly into existing technologies and your lean product development system before you use it.
2. Use the technology to support your lean development process, not vice versa.
3. Technology should enhance people, not replace them.
4. Do not look for silver bullets to enhance your product development performance—there are no short cuts. There are no ways to avoid the hard work required to develop a high-performance product development system.

Remember that any one of your competitors can buy or develop the same technologies. You get the most out of technology when it not only fits within your system, but when it enhances your processes and promotes continuous improvement—the human is still responsible for getting the work done.

CHAPTER 14

Align Your Organization Through Simple, Visual Communication

"In business, excess information must be suppressed.
Toyota suppresses it by letting the products
being produced carry the information."

TAIICHI OHNO

MANY OF THE PROBLEMS COMPANIES had in the "bad old days" before concurrent engineering were the result of the "chimney" phenomenon. Specialists lived in chimneys and communicated upward and with other specialists, but poorly across chimneys. Today, if you ask a room full of engineers if communication is important to effective product development, you are likely to get amused shrugs—the answer is obvious. After all, product development is information flow among many specialists. Stop communication, stop information flow, and you stop product development. Now, instead of "throwing the design over the wall," engineers are taught to communicate concurrently with a team of upstream and downstream specialists—across functions. Almost everyone understands that more communication is better and that collocating engineers in the same office area so they can communicate intensely every day is a PD "best practice."

Given that everyone agrees that communication is critical to good product development, what is left to say on this subject? Actually, quite a bit, including the fact that more communication is not necessarily better. And that sometimes face-to-face communication is not as good as written documents. And that large-scale collocation may not necessarily be all it's cracked up to be.

If this is surprising, just think for a moment about endless concurrent engineering meetings where time is lost just getting different functions to agree on basic terminology and how unproductive this "intense" communication can be. What you really want is selective communication that gets the right information to the right people at the right time, so that they can make good decisions. Inundating people with too much information often means they cannot sort out what is core from what is peripheral. It is inefficient, ineffective, exhausting, and wastes time and talent, and is sometimes less productive than the common practice of the bad old days of throwing drawings over the wall and saying nothing else. Somewhere between these two constructs is a happy medium that avoids information overload but gets necessary information to others effectively and efficiently.

Chapter 4 described how Toyota front-loads the product development process. The essence of simultaneous engineering is bringing downstream considerations to the table early in the development process, when options are the most fluid. Set-based concurrent engineering was discussed as a way to consider many options in this fluid stage and then narrow the range simultaneously across functions. Obviously, communication is key to doing this, but this means streamlined communication that makes every note, chart, meeting, and report count. This chapter provides an overview of the tools that Toyota uses to keep communication highly focused, simple, and visual in a lean PD system.

Chief Engineer's Concept Paper: An Aligning Document

Chapter 7 discussed the central role of the chief engineer and Chapter 3 discussed how the CE's concept paper launches and defines the core parameters of the entire product development proposal at Toyota. The concept paper is highly confidential, typically runs 15 to 25 pages in length, and text is supplemented by tables, graphs, and sketches intended to provide the team with a single unifying direction and decision-making guideline. With this document, the CE aligns many functional specialists efficiently toward a common vision.

This is not to say that the chief engineer creates this document in a vacuum. Because Toyota's PD process encourages the interchange of many ideas to get the best ideas on the table, the CE pulls the vehicle concept together by drawing on diverse inputs (from planning, styling, marketing, purchasing, etc.). To integrate these, the CE holds many meetings and often presides over heated discussion and debate. The CE examines the customer data provided to him by the marketing department. Practicing *genchi genbutsu*, he or she also observes and learns about the day-to-day activities of his target audience first-hand. (Chapter 3, for example, mentioned the *Sienna* minivan chief engineer who personally drove through every U.S. state, every Canadian province, and every Mexican state to experience firsthand the many driving conditions of North America.)

In addition to *genchi genbutsu,* the chief engineer relies on *nemawashi,* getting proposals and reactions from many people and modifying the collective input to reflect a consensus. *Nemawashi* is not peculiar to Toyota (it is a staple of Japanese management), but it is central to any problem solving at Toyota. In fact, a common refrain of Toyota CEs is "My power comes only from personal persuasion." No final proposal goes to executive decision makers without broad circulation. So that essentially, the proposal is already approved by the time of the formal meeting. Once the CE has a vision for the new vehicle, he or she meets with both internal and external technical specialists to identify subsystems that might be right for the vehicle. This group of specialists often includes representatives from suppliers who have developed a particular technology (such as in seating or lighting) that might be a good fit for the vehicle vision. The CE works to identify technologies that are right for the vehicle and then determines, with the help of the specialist(s), which of the required technologies are already fully developed and financially feasible. The CE team will invite suppliers whose technologies pass the CE's criteria to participate in technology demonstrations for further consideration. The CE will also consult product engineering specialists or visit manufacturing plants to get input on the manufacturability of particular technologies or styling characteristics that are under consideration. All of this ensures that the concept paper will be viable and fully grounded in data and current capabilities. Once the document is completed, the CE distributes the paper to all assistant CEs and functional staff leaders on the program.

The way the concept paper is conceived and delivered illustrates the importance of having a clear delineation of roles and responsibilities. It lays the groundwork for knowing what to communicate and what not to communicate. The lean view of communication is:

- If everyone is responsible, no one is responsible.
- If everyone must understand everything, no one will understand anything very deeply.
- If all communication is going to everyone, no one will focus on the most critical communication for their role and responsibility.
- If you inundate your people with reams of data, no one will read it.

At Toyota, while there is a great deal of information shared across functions, it is targeted information. When a prototype of a vehicle is being reviewed, each engineer represents a function. They come prepared with checklists for their respective function to evaluate all critical points.

Each engineer is intensely focused on his or her piece of the vehicle and any interfacing components. Any suggestions go directly to the engineer responsible for that part of the vehicle. Moreover, the responsible individual must consider the problem or suggestion and come up with a response (though not necessarily on the spot). The individual writes comments, concerns, or suggestions on a flip chart for that particular part of the vehicle. He or she later fills out a wall chart stating the problem, the proposed countermeasure, and when the action will occur.

The Cross-Functional *Obeya*

When cross-functional communication became a standard at Toyota, it was originally handled through targeted meetings or reports and the CE's integrating role. But the practice took a major leap forward with the development of the *obeya* or "big room." As discussed in Chapter 7, CE Uchiyamada used the *obeya* for the development of the *Prius* hybrid vehicle as a way to compensate for his lack of experience. Whereas previous Toyota CEs had engaged in weeks of discussions with individual engineers, one by one, before making decisions, Uchiyamada discovered that despite his lack of in-depth product knowledge, he was able to make decisions in a single brief meeting with engineering managers representing different functional departments.

Uchiyamada's decision supports the views of advocates of concurrent engineering (simultaneous engineering) who have found that collocation greatly intensified communication and led to solving problems early, before a lot of capital commitment. But, as mentioned in Chapter 8, there are some rather unique features of the *obeya*:

- *Engineers are not collocated.* It is the engineering staff leaders (leaders of each functional group) who meet with the chief engineer regularly. These engineering managers have desks in their functional organizations and come to the *obeya* regularly (often daily) for long, collaborative work sessions (not traditional meetings).
- *Visual management is key to effective communication.* Engineers plaster the room's walls and "mobile walls" with information organized by vehicle part. The functional head responsible for that part of the vehicle is responsible for posting and maintaining the information, and this information allows anyone "walking the walls" to assess program status (quality, timing, function, weight) up to the day.

- *The* obeya *has evolved through careful PDCA.* One of the innovations the PDCA process has produced is that, as a program progresses, moving from engineering to prototyping in engineering facilities to preparation of production in the plant, the *obeya* is also moved. To accommodate this, the "traveling *obeya*" was developed to move downstream to the plant as the program moves downstream. An additional modification was to incorporate the module development teams into the *obeya* process. Each modification in the *obeya* has been deliberate, studied, adjusted, and then standardized across programs.

The *obeya* has been a critical part of Toyota's success in radically reducing lead time, but part of that success was that Toyota did not disband the functional organization or collocate hundreds of engineers. The *obeya* system truly combines the best of both the latest technology and Toyota's traditional lean practices. The *obeya* contains state-of-the-art CAD computer systems and digital projection technology to enhance real-time viewing of design, simulation, and test results to foster cross-functional collaboration. However, paper visual management continues to be a centerpiece of program management, but paper is now hung on the walls of the *obeya*. Functional managers are responsible for information in their areas of specialty and in typical meetings present the status of their work referring to the visuals on the wall. Through regularly scheduled (but selective) cross-functional meetings of key leaders, Toyota has improved the functional organization by facilitating quick decision making.

Alignment Tools

Any company that employs 500 to 1,000 technical professionals working on pieces of a complex system and several hundred suppliers developing and testing components must have a system for aligning these individuals and their work. *Alignment means you harmoniously bring together all the individual inputs from various people at the right time to achieve the desired objective.* There are a limited number of ways to make this happen:

1. *Individual level.* Each person can operate independently. If the technical requirements are crystal clear and if all these individuals have the skill and knowledge to do exactly what is required, they can join

their separate inputs and it will all fit together. This would require very detailed and unchanging technical requirements.

2. *Team level.* One big team can hash out each and every detail in meetings or virtual meetings to come to agreement each step of the way. This is obviously time consuming.

3. *System and subsystem level.* This means dividing an overall system into relatively autonomous subsystems. The subsystem teams, following clear standards, can work separately to ensure integration within each subsystem and to meet the overall requirements necessary for the subsystems to work together.

4. *Horizontal integration.* Different individuals and subsystems work separately while an "integrating force" ensures the work is well coordinated across parts of the system. This type of integration is best facilitated by a super integrator who directs a number of people in liaison roles or by cross-functional task groups.

Toyota uses all of these methods in varying degrees. The chief engineer plays a super integrator role while module development teams manage integration within subsystems. Driven by clear targets starting with the chief engineer concept paper, standardized skills and checklists enable individuals to work separately. A variety of meetings brings people together to focus on integration issues. In addition, Toyota uses a specific set of tools to aid in the integration task, some of which are discussed below.

Nemawashi at Toyota

Nemawashi is not a tool in the traditional sense, but it is a key to integration and underlies the use of many other tools, such as A3 reporting described later in this chapter. In Japanese, the word *nemawashi* originally referred to making the preparations necessary for transplanting a large tree. In the business world, *nemawashi* is about achieving consensus among stakeholders prior to the actual formal decision-making event. *Nemawashi* in product development usually involves communicating relevant data or information to the appropriate PD team members and holding small, informal, usually technical, discussions about potential solutions to design or manufacturing challenges. This often takes place outside of formal meetings so that meeting time is used primarily to inform upper level managers of the options and the group's recommendation. An A3 or decision matrix typically communicates this.

If the design or manufacturing challenge requires a major change in process, in design standards, or significant funding, then a more formal decision-making process will be used. All engineers, early in their careers, learn the *nemawashi* process—share the problem with others as you work on it, get others' input, and buy in so the final product is better and has already been signed off as it was being developed. You never want to be in the position of defending your own solution, even if it is technically strong and defensible.

The *Ringi* System at Toyota

The *ringi system* is a more formal decision-making process used for handling significant decisions. In the *ringi* process, a small team of people with the necessary expertise is assigned to analyze some specific issue or challenge and recommend a solution. At the conclusion of the analysis process, the team creates a decision document called a *Ringi-sho*, which outlines the challenges, the countermeasure, and the potential implications, both positive and negative, of adopting the proposal. The team then meets with all managers who will be affected by the proposal and requests their approval. Sign off on the proposal is traditionally done with a manager's *hanko*, a personal stamp used only by managers at a certain level. Originally, Toyota required a minimum of eight *hanko* for proposal approval. However, in 1990 a group within Toyota initiated the "three-stamp campaign" to reduce the number of approval stamps. While this narrowed the number of people responsible for maintaining essential alignment, it sped up decision making.

In product development, *ringi* typically takes place early in the process during the *kentou* or study phase. Small cross-functional teams are given a limited time period (approximately 30 to 90 days) in which to analyze a challenge, meet with managers, and diffuse the document throughout any effected product development organization. In product development, *ringi* can be used when a novel product design requires a significant change in the standard manufacturing process or production engineering wants to try a new, potentially more efficient, nonstandard manufacturing process. In both cases, product engineering, production engineering, and the manufacturing plant would have to sign off on the *ringi-sho*.

Using *ringi* in product development has several advantages. It gathers input from all effected parties and aligns effected groups, reducing decision making to a single step during the product development process. In addition, it provides a constructive outlet for nonstandard approaches while simultaneously providing organizational support for standardized designs and processes. The underlying message is to follow standards unless there is a very good reason not to—good enough to go through the *ringi* process and prove it.

Hoshin Management at Toyota

Hoshin management (a.k.a. policy deployment) is an effective tool for aligning an organization toward the achievement of broader goals or objectives and allowing that organization to react quickly to a changing environment. In a lean organization, *hoshin kanri* is generally an annual planning tool that aligns the organization's long-term vision with its shorter-term activities while also aligning the efforts of people in the organization with the goals of the organization. (See Figure 14-1.)

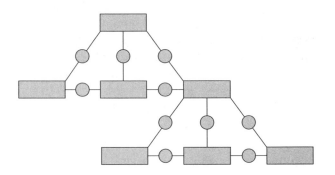

Hoshin management aligns the activities of people throughout the company so that the company can achieve key goals and react quickly to a changing environment.

(Shiba, Graham, and Walden, 1993)

Figure 14-1. *Hoshin* Management is an Alignment Tool

There are six primary components to *hoshin*. The first of these is company vision—the long-term vision of where the company wants to go, what it is, and what it wants to become. The annual *hoshin* complements company values. In connection with this, *hoshin* also considers specific annual goals, the means to accomplish these goals, metrics to measure progress, target value for the metric, and deadline and follow up. There are four phases to *hoshin* at Toyota.

1. *Strategic planning.* Strategic planning has different levels. In a broad sense, it can mean developing a ten-year strategic plan identifying the specific target market for the entire company. *Hoshin* is not used for this except as a vehicle for considering a set of short-term steps to achieve the long-term vision. Like other strategic planning approaches, *hoshin* identifies problems and opportunities based on company performance and other environmental data (including shortcomings from previous *hoshin*). This, in turn, can be used to develops a near-term, future-state vision based on these factors.

2. *Hoshin deployment. Deployment statements* must be based on facts and data, be operationally viable, and be aligned with the six primary *hoshin* components. At the Toyota Technical Center in Ann Arbor (TTC) and at all Toyota facilities, the president's *hoshin* event kicks off the year. The president communicates his *hoshin*, based on the chairman's *hoshin*, to TTC employees (see Figure 14-2). These statements are communicated to the various functional groups and each group must translate the *hoshin* into a meaningful goal for that function. For example, the president may make a *hoshin* of reducing product development lead time by 20 percent by employing new methods and tools. The vice president of Production Engineering might then, in support, make a *hoshin* of reducing Production Engineering lead time by 20 percent, by more fully leveraging solids die design capabilities and an automated pattern machining project by the "X" program. The die design manager then in turn develops several supporting *hoshin*, such as training personnel, working with the technology supplier to fully adapt the technology to fit the Toyota process, and piloting the technology on specific applications. At each level of deployment, the participants go through a goal-negotiation process called *catchball.* This *catchball* process allows participants to develop a clear understanding of the *hoshin* and enables the *hoshin* recipient to "enroll" in the objectives.

3. *Controlling through metrics.* Both results and means are measured and reviewed on a regular basis in a PDCA methodology. In the above example, TTC would measure both the reduction in

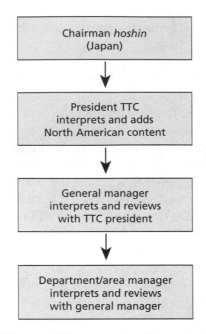

Note: Also done for all Toyota affiliate companies. (Toyota Auto Body, etc.)

Figure 14-2. *Hoshin* Example at Toyota Technical Center (TTC)

lead time (results) and the implementation of specific enablers (means), such as solids die design or the automated pattern process. This provides insight into progress in implementing the means and whether or not your means are having their intended results.

4. *Check and act.* During the year, managers measure actual progress against the goal and make necessary adjustments. In most cases, managers are required to develop a fallback plan that they can implement if the difference between the goal and reality is too great.

A key to using *hoshin* effectively is to close the loop through rigorous reviews at each level (see Figure 14-3). Each pair of levels must first agree on the critical few objectives for the year and how they will be measured. Then supervisor-team associate pairs meet periodically to review progress toward the *hoshin*. At TTC, this occurs three times per year and each individual is evaluated against specific *hoshin* targets. Although problems are addressed during this performance review, the meeting focuses on the

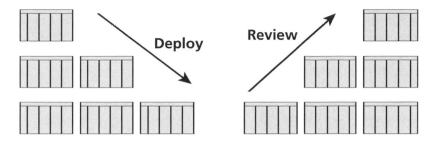

Figure 14-3. *Hoshin* Deployment and Review Process

hoshin—the specific objectives of the individual to meet the high-level strategic objectives.

An interesting example of strategic use of *hoshin* in product development was the chairman's *hoshin* at the beginning of the twenty-first century. Based on Toyota's commitment to the environment, the chairman requested an increase in average fuel economy through the use of new technologies. The product development community responded with, among other things, the *Echo*. Commitment to the environment is part of Toyota's mission statement and a core strategy to position the company for the future. This *hoshin* launched many concrete initiatives to make a big leap forward toward that vision.

Toyota's A3 Problem-Solving Tool

A3 refers to a standardized communication format, a disciplined process of expressing complex thoughts accurately on a single sheet of paper. As the following comments made by Toyota managers regarding the A3 report show, it is much more than that:

> "Force yourself to filter and refine your thoughts to fit one sheet of paper in such a way that management has all of their questions answered by reading a single piece of paper—it is the essence of lean."
>
> "A3 is much more about disciplined thinking than it is about any particular writing technique."
>
> "Imagine how you would feel if you wanted to give me a report, but I insisted that you could only draw a single picture to communicate your information?"

A3 is a standardized technical writing methodology to create a report on one side of a standard size piece of paper to guide problem solving and achieve clear communication across functional specialties. There are four different types of A3 forms (see Figure 14-4).

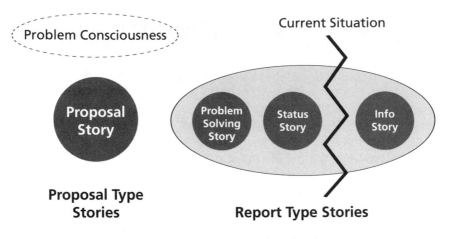

Figure 14-4. Four Types of A3 Stories

1. *Proposal story.* Used for proposing a plan or new initiative, this story always revolves around a theme. Although it does not end with an actual proposal, it requires a clear plan, identification of issues that will have to be resolved, and a schedule. (See Figure 14-5.)
2. *Status story.* Used for giving the status of an ongoing initiative, this story calls for actionable information (see Figure 14-6, page 270). What was the objective of the project and how is it going relative to the objective? What was the planned versus actual implementation time line? What issues need to be resolved, and what future actions are planned?
3. *Informational story.* Used to share information, for example, about a development at a competing company or elsewhere in Toyota, informational A3s are "free-form" and the layout and delivery of the story is left to the writer to formulate.
4. *Problem solving story.* Used when a plan, goal, or standard exists and the company is not meeting it (see Figure 14-7, page 271). That is, there must be a problem. The problem-solving story needs to be thorough, communicating the complete process of plan-do-check-action. Embedded in this storytelling is Toyota's approach to "practical problem solving," making clear the goal, the data on the current

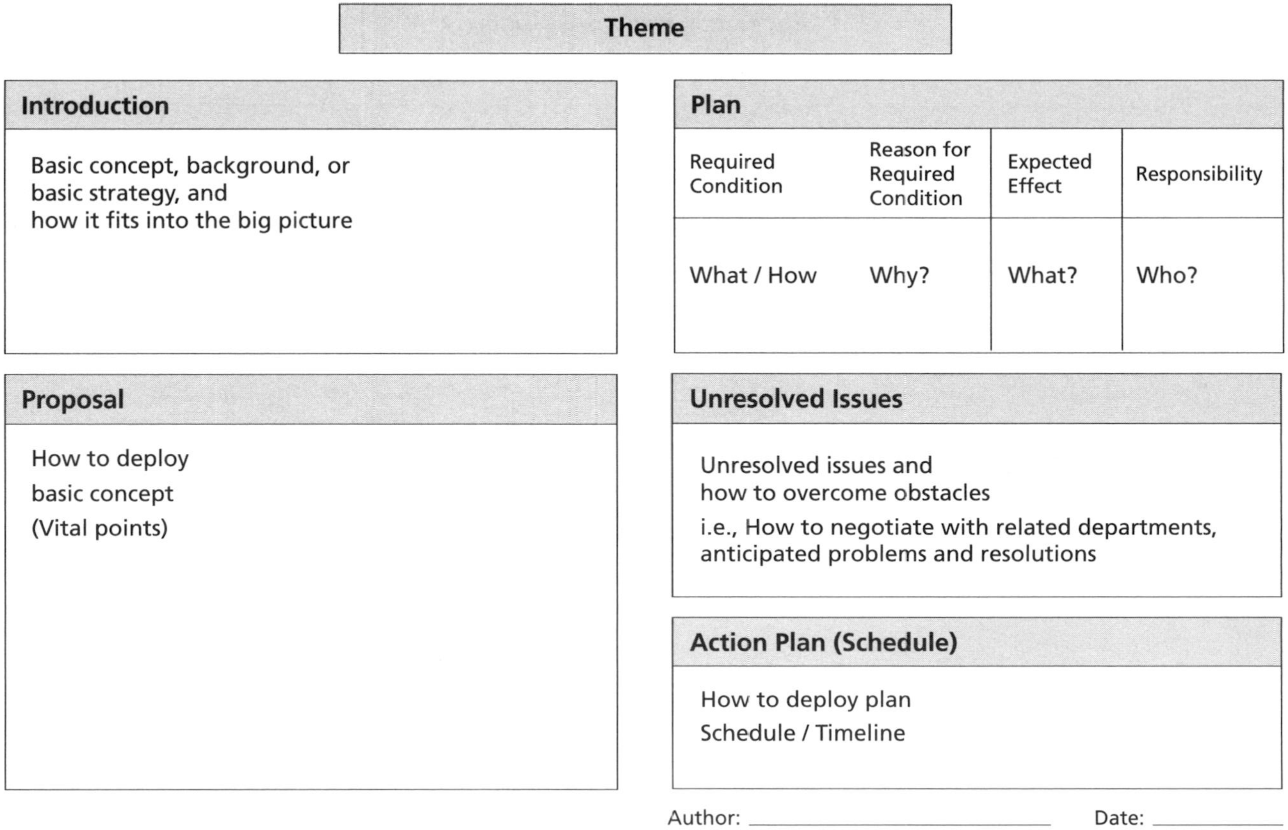

Theme

Introduction

Basic concept, background, or
basic strategy, and
how it fits into the big picture

Plan

Required Condition	Reason for Required Condition	Expected Effect	Responsibility
What / How	Why?	What?	Who?

Proposal

How to deploy
basic concept
(Vital points)

Unresolved Issues

Unresolved issues and
how to overcome obstacles
i.e., How to negotiate with related departments,
anticipated problems and resolutions

Action Plan (Schedule)

How to deploy plan
Schedule / Timeline

Author: _____ Date: _____

Figure 14-5. Proposal Story Example

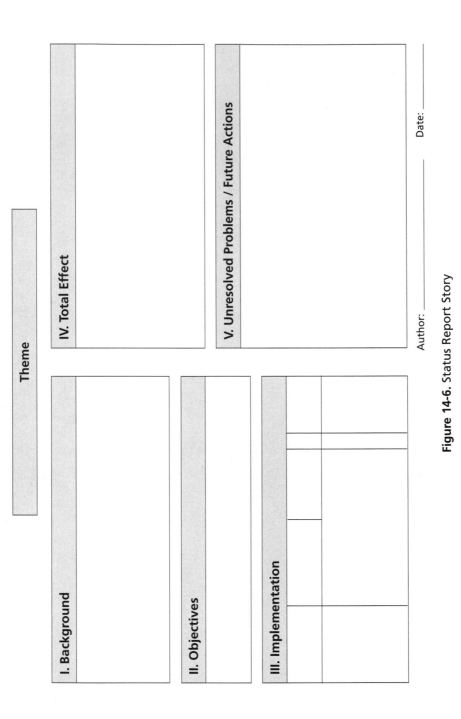

Figure 14-6. Status Report Story

Theme

Answer the question: "What are we trying to do?"

Problem Situation

- The standard
- Current situation
- Discrepancy / Extent of the problem

Rationale for picking up the problem
(Importance to business activity, goals, or values of the organization)

Target / Goal

Measurable description of what you want to change; quantity, time

Cause Analysis

Problem:

Potential causes

Most likely direct cause:
Why? → Why? → Why? → Why?

Root cause:

Countermeasures

(Resulting from cause analysis)
- Temporary measure
- Long-term countermeasure

Implementation

What	Where	Who	When
Actions to be taken	Location of activity	Responsible person	Times, dates

Follow-Up

- Unresolved issues and actions to address them
- How will you check effects?
- When will you check effects?
- How will you report findings?
- When will you report findings?

Author: _____ Date: _____

Figure 14-7. Problem Solving Report, Proposal Story Example

situation, a detailed analysis of the root cause, countermeasures, and the story of implementation. The follow-up must include the verification (check) that the countermeasure worked, future actions to use, and what was learned in the verification process. An example of a problem-solving story for a safety issue is shown in Figure 14-8.

Although an employee's immediate supervisor is the primary teacher of A3 writing, Toyota offers universal "vital points" for both A3 writing and problem solving during training, recommending specific uses for graphs and other communication aids. Below are some vital points for problem solving and for writing A3s.

Problem Solving Vital Points

1. Grasp the situation based on facts
2. Observe the problem firsthand
3. Address a single abnormal occurrence (one A3—one problem)
4. Observe the abnormal occurrence at the point of cause
5. Investigate the cause thoroughly and review all facts and data
6. Use temporary (containment) measures when necessary
7. Identify the root cause
8. Develop countermeasures and assignments, and set deadlines

---

A3 Writing Vital Points

1. Plan time to grasp the ENTIRE situation.
   a) Consider a wide range of information sources
   b) Base story on facts not opinions
   c) Consider the long-term effect
2. Decide what kind of story you need to tell. Write to your audience. Consider their needs and knowledge of the situation.
3. Relate the story to company values and philosophy.
4. Make the story flow in logical sequence.
5. Save words. Use graphs and visuals to tell your story whenever possible.
6. Make every word count. Be specific and avoid jargon.
7. Consider the visual effect of each box on the page in helping you tell your story.

TO: _____

PROBLEM SITUATION

REDUCE INJURIES DUE TO CUTS DURING THE HANDLING OF SHEET METAL

Injury Type	Hand Injury descriptions	# of incidents	# of incidents resulting in days off	Average # of man-hours lost per incident	# of incidents resulting in lost days
Hand	Minor scrapes and cuts from sheet metal handling requiring in-house first aid treatment	6	1	19.7	60
	Major cuts from sheet metal handling requiring stitches or other medical treatment	10	10	37.6	60
Back		10	10	15.4	4
Eye				17.5	4
Arm		1	1	19.3	8
Leg				6.3	4
Foot				24.6	2
Head				6.0	3
Hernia				0	3
Chest				0	2
Lungs		1	1	0	1
Neck					

Note: Major injuries are recorded to a maximum of 2 weeks of lost time (i.e. broken bones, injuries requiring long periods of disability, death and dismemberment)

Total (plant-wide) lost man-hours due to injury for 2001 is unacceptable → 1550 hrs
Company objective is to reduce total by 50%. (1550 hrs → 775 hrs)

Significant hand injuries (requiring days off) are primarily from handling materials in an unsafe manner. Most injuries are occurring in Press/Fab/Assy areas where material handling is required as part of the job.

350 hrs of lost time were recorded in 2001 due to hand injuries. Hand injuries costs Goliath an estimated $14,200 per year in lost man-hours.

Employee safety is one of Goliath's key company values and must be addressed.

TARGET/GOAL

Reduce sheet metal handling injury frequency by 90% over the next 12 months.

CAUSE ANALYSIS

PROBLEM: Employees are receiving cuts, scrapes, and abrasions while handling sheet metal.
MOST LIKELY CAUSE: Employees are not following "gloves required" policy when handling sheet metal parts or blanks.

WHY? For small or quick jobs when gloves are not handy, employees would rather risk getting a cut then expending the required effort to find a set of gloves to put on.
WHY? Lack of discipline to company policy
WHY? Human nature to take the easy route – perceived benefits outweighs the risk
WHY? No motivation to follow rules when it is not convenient to do so
WHY? Penalties for breaking rules are not being enforced AND/OR lack of sufficient reward for adherence to policy

ROOT CAUSE: Motivational issue → Employees are not motivated enough to expend the required effort to follow basic shop safety requirements when it is inconvenient to do so.

COUNTERMEASURES

Clarify definitions and conditions for applying shop safety rules with Union representatives and shop supervisors. Rules may need to be reworded and reworked to reflect practical shop floor application.

Reward system will be implemented as a first step in lieu of increasing employee disciplinary action for failing to follow company safety rules.

Raffle consisting of a cash prize (suggested value of at least $2,000) will be held at the end of the year. To maintain eligibility, shop floor members must:
• Maintain a clean personal injury record
• Not be caught failing to follow shop material handling and eye protection safety requirements

Employees would be encouraged to inform and watch out for each other throughout the workday. Once or twice a week, a randomly selected member of the supervisory staff would perform a 'shop patrol' walk to look for employee non-conformances.
Eliminated employees would be given the option to buy back into the raffle by making a minimum cash donation to a charity (to be determined).

Subsequent safety infractions after being removed from raffle eligibility may result in an employee write up.

IMPLEMENTATION

To be implemented as a company safety initiative in conjunction with Union Plant Safety Committee Tracking is to begin for the abbreviated year, starting in March 2002.

ACTION REQ'D	RESPONSIBILITY	DUE BY
Project approval	President (Lowery)	Feb 8
Communicate A3 plan to Union Safety Committee for review, discussion, and roll out strategy	HR Employee Relations (Eizerman)	Feb 18
Clarify shop safety rules	Goliath-Union Safety Subcommittee	Feb 25
Roll out details to Goliath managers and supervisors + shop employees (through team meetings)	Goliath Safety Representative (Ganc)	Feb 28

VERIFICATION and FOLLOW UP ACTIVITIES

Progress to be tracked monthly during Quality Systems Team meetings measurables tracking (Compare 2002 progress with '99 / '00 / '01 YTD safety data)

Informal survey of shop supervisory and managerial staff on a quarterly basis regarding shop safety improvement and compliance to shop safety rules.

AUTHOR: _____
DATE: _____

Figure 14-8. Hand Injury Reduction A3 Solving Report

There is an interesting parallel between the way Toyota views the use of space in its Toyota Production System and its A3 reports. In a lean factory, it is very bad to fill space with *muda* or waste, which is generally excess inventory. High value-added floor space is very valued. "A lot of extra stuff" lying around leads to disorganization, and it is difficult to distinguish the correct standard condition from the out-of-standard condition. Likewise, in A3 reporting, the point is to have high value-added documents in which it is easy to see the critical points unimpaired by waste—too many words, elaborate explanations, graphs, etc. *Muda* in documents obscures the message and often causes people to leave out key points.

Toyota's A3 reports and the discipline with which they are prepared provide fast and accurate communication that the whole organization understands—crucial for a lean PD system. The A3 also facilitates continuous improvement. What the A3 report requires gets done, for example, the know-how database will be updated to reflect any change in a standard or addition to knowledge that resulted from the A3 process. In lean thinking, the focus is on learning from problem solving (not just fixing the present situation), and the A3 is a tool that enables that effort.

Communication and Alignment at Toyota

Many companies have tried to implement similar communication methods and alignment strategies. But whether they use management by objectives, project management, stage-gate models, consensus decision making, collocation, resident engineers, cross-functional teams, or even A3 reporting, few seem to do this as effectively as Toyota.

The probable cause for this distinction is the way a company assigns roles and responsibilities, follows through on managing these, and whether it provides a minimum degree of organizational stability needed to make these tools work. Toyota, which has a very strong functional organization with across-functional integration, through the chief engineer system excels at cross-functional communication because it provides a working environment that is stable, standardized, and continuously improving. This organizational stability starts at the leadership level, which provides consistent direction about what is important. Stability at the working level includes spending time within a functional specialty and gaining in-depth knowledge. Stability also enhances the ability to learn from program to program.

Roles and responsibilities mean that there is always an individual responsible for carrying through the actions. In other companies, the pendulum too often swings either completely to individual accountability or to teams, teams, teams. Toyota has found a healthy balance. Though the A3 report is actually a group activity and *nemawashi* is critical to building consensus in the A3 process, an individual author signs the report. If there are coauthors, all share in the responsibility. Thus the A3 process is an alignment tool that manages and make roles and responsibilities work. In the communication and alignment process in lean PD, there is an important interplay between teamwork and individual work. Individuals always do detailed work. Teams provide ideas and input, share in decision making, support each other, and share in implementation.

Ultimately, alignment and effective communication give a company the ability to develop standards throughout the organization. People critically evaluate and improve on these standards, creating revised standards as needed. This is, in fact, the backbone of learning as an organization. The next chapter which covers the 13th and final LPDS principle, *use powerful tools for standardization and organizational learning*, revisits this concept.

LPDS Basics

Align your organization through simple, visual communication

It is fairly obvious that communication is a good thing in product development. You cannot develop new products without good communication. But LPDS recognizes that you can get too much of a good thing. Communication should be targeted, sufficient, accurate, and focus on the essential facts. Toyota has developed a number of mechanisms to aid effective communication. The CE paper provides important program direction and creates alignment. The obeya system is a mechanism for on-going cross-functional communication and status tracking. *Nemawashi*, the *ringi* system and *hoshin* provide tools for organizational alignment. The A3 process helps to bring discipline and standardization to improve the effectiveness of all types of communication—especially communication about problem solving.

CHAPTER 15

Use Powerful Tools for Standardization and Organizational Learning

"Engineers at the Toyota Technical Center use hetakuso-sekke, *which is the small booklet containing the failures experienced in the past."*

KUNIHIKO MASAKI,
former President of the Toyota Technical Center, Ann Arbor

COUNTLESS COMPANIES HAVE SPENT untold hundreds of millions of dollars chasing after the performance improvements promised by organizational learning and effective knowledge management. They have made mega-investments in the information revolution, pouring cash into networked knowledge, massive online databases and all of the latest hardware and software. Yet despite these colossal investments of both human and capital resources, most companies have continued to struggle to make organizational learning a true competitive advantage. One reason is because their focus has typically been on tools that manage explicit or procedural knowledge, which is, although expensive, relatively easy to do, and easy for competitors to replicate. As we have previously discussed, it is in leveraging "tacit" knowledge, or know-how that is the greatest source of competitive advantage. The focus must be on tools that help the organization change the way things actually get done. This type of knowledge is embedded in people and culture; it is around tacit knowledge that the best learning tools and technologies are built and designed.

How Does Your Organization Learn?

Chapter 11 identified *organizational learning* as a key source of Toyota's competitive advantage. But just how are they able to succeed where so many other companies have struggled? What tools and methods do they employ and how do those tools differ from what their competitors use?

One way to understand Toyota's tools and methods is to once again study a contrastive parallel to NAC. Although staffed with many capable engineers, NAC struggled to recreate itself as a learning organization. Like Toyota, NAC recognizes the importance of learning, and many individuals within the company have learned well through trial and error. These individuals grew increasingly more valuable over time. Unfortunately there was no mechanism within NAC to spread what had been learned across the organization and to the next generation of engineers—at least not to a great extent. And if a veteran employee left the company, they generally took years of accumulated wisdom with them.

A lean PD system thrives best in a company that understands and cultivates organizational learning, develops systems, and evolves a culture that captures tacit knowledge and turns it into standards that everyone learns and passes down to others. The importance of learning is embodied in many of the tools already discussed in this book. We have already discussed a number of Toyota's most effective learning tools (technical mentoring, *hansei*, A3) in great detail, and we will not revisit them in this chapter. This chapter focuses on a number of other lean PD tools that help Toyota's learning capability: know-how databases, checklists, competitive teardowns, *senzu*, process sheets, and quality matrices. The power and effectiveness of these tools is best illustrated by comparing how they are used within an organization like NAC and within a learning organization like Toyota.

Knowledge Database at NAC: The Body Development Value Stream

Although NAC made multiple attempts to create knowledge bases to guide and assist engineers, there was no central strategy for knowledge and information management along the body development value stream. Instead, many individual functions and locations developed their own version of a knowledge database and, of course, caused informational inconsistency, even outright contradictions, across those databases. What undermined this system even further was that the primary specialists are seldom involved in the database creation, validation, or maintenance. In general, these specialists had little faith in the information, a prerequisite for any knowledge tool to work. Consequently, what information was available was seldom used. To exacerbate this condition NAC did not have the fundamental discipline, values, and reward system in place prior to creating the database, and functional specialists did not view it as a particularly useful tool.

NAC has also invested in a *knowledge-based engineering* pilot. This rule-based "intelligent" technology can continually accumulate and codify new information and would utilize the latest data to actively guide engineers in their design process, literally locking out certain choices and providing automatic design sequences. Developed and maintained by IT staffs, and in some cases, outside contractors, who work to extract information from engineers and download it into databases, this technology has not only failed to yield the promised benefits, but it is, in our view, potentially harmful. It has not worked because it has not earned credibility with the engineers; they have not been enrolled in the development process. It has not worked because it violates one of the primary principles of lean technology deployment; it seeks to replace, not enhance its userbase. And it is potentially harmful because it removes the engineer as an active participant in the learning process.

The Know-how Database at Toyota

The compilation of the many standards checklists Toyota has used for over 40 years is the basis of its computerized know-how database. It is important to understand that these years of accumulated experience worked very effectively in paper form before they were ever computerized. For example, engineering checklists were organized in notebooks that were kept by individual functional groups. Each functional group had its own notebooks (e.g., 3-ring binders). At the beginning of a program, the production engineering group responsible for doors would share its most updated door checklists with the body engineers responsible for doors. Since Toyota has such a clear organization of functional roles and responsibilities it is not necessary to send all of the checklist information every which way. It is clear which specialists are working on which parts and only they need the information in notebooks for their parts of the vehicle. The notebooks were not profound and did not replace deep engineering knowledge. They just remind the engineer to think of each aspect: Did you check whether two parts are interfering with each other or not? Did you check that the gap conforms to standards? Does this ratio fall within a standard range? There may be a graph showing not to exceed this level. In each case, the engineer physically makes a check to note I thought of that or did that. It is much like a pilot's flight checklist. It does not make the pilot a great pilot, but it can help avoid basic mistakes.

The *hetakuso-sekke* described by Masaki-san in the opening quote is a different type of standard reference guide mainly used in body engineering. In this case Toyota put together a set of rules of what makes a good and bad design mainly through diagrams. For example, in body engineering many had to do with how to reduce weight. For example, there may be a plate in which a hole can be made for weight reduction. When there is a flange a corner radius can be used to minimize weight. It is considered by senior engineers to be a "lightweight concept." In fact, the name means something like foolish engineering or shallow thinking designing. An experienced engineer should not need such tips. But for a younger engineer it can be a very valuable source of design tips.

Putting these drawing tips and the checklists on paper in the old days was a useful way to maintain a knowledge database. But it was not the only knowledge database. The real database resided in the heads of the experienced engineers with "towering technical competence." Each of these engineers is a living database with extraordinary knowledge in a very specific area. If you know who to talk to, or as long as the right people review the design, that knowledge database is activated. So simple checklists or booklets are supplementary.

Putting these written forms of knowledge onto a computer does not replace the deep knowledge of functional experts. And the written checklists did their job quite well for many decades. But a computerized knowhow database is still a significant advance. Engineering guidelines, graphics, and practices are outlined and explained in greater detail than was possible on handwritten checklists. Information is arranged by vehicle/part or just part type, and within the part-type database, there is a competitor benchmark component that allows engineers to view pictures of competitor products and teardown analysis. Other data are organized by manufacturing process/part type and include access to the quality matrices and *senzu* discussed later in this chapter. In addition, the knowhow database is linked to the design database to facilitate importing and exporting information/geometry. There are also manufacturing process sheets, complete with descriptions, pictures, and even videos of processes occurring right on the factory floor. Users can also access quality and performance data by part by plant.

One potentially big step forward in computerizing the checklists is turning them from simple rules of what to avoid or what numerical values to use to explain the reasoning behind the rule. An engineer who sees the old checklist has no way of knowing from the checklist the reason behind the rule. It provides know-what but not know-why. The new know-how databases are evolving to explain the reason why. This is not as important when the experts who created the rules are right in the room with you in Japan. But as Toyota is globalizing product development it is not as easy to get access to the true experts, and it is important to include the reasoning process so the engineer can think through the problem instead of simply following a rule.

A key to the success of the know-how database at Toyota is seamlessly integrating it into the larger system, V-comm. V-comm is a fully integrated network of tools that links active design, simulation, test, and communication technologies along with the know-how database to provide engineers with a powerful suite of tools to enhance their productivity. By integrating the know-how database into the basic engineering tool set Toyota emphasizes that learning is not an add-on or "extracurricular," it is a basic part of the job.

This database has been successful for a number of reasons. As previously noted, the underlying company values, discipline, and reward for adhering to the process already exist, so people automatically use and learn from this information. Moreover, the database is centralized; there are no competing knowledge databases to create confusion or contradictions. Functional specialists maintain, verify, and update their own portion of the database, just as they did with their handwritten checklists. This typically takes place after a *hansei* event (a reflection or learning event discussed in Chapter 11). Finally, the database is designed to enhance the performance of Toyota's people, not replace them or withhold the design process from them. In a lean PD process, individual engineers are central; learning tools and technology should reflect this.

Communicating and Evaluating Sets

As we have previously discussed, set-based concurrent engineering (SBCE) is a cornerstone of Toyota's product development process. Moreover SBCE is a powerful source of knowledge and continuous improvement. However, communicating, evaluating, and learning effectively from numerous alternatives, each possessing diverse and technical design characteristics can be challenging. Consequently Toyota engineers have learned to utilize two important tools to aid in this task.

Trade-Off Curves

A trade-off curve is a relatively simple tool that is consistently used by Toyota engineers to understand the relationship of various design characteristics to each other. In a trade-off curve a subsystem's performance on one characteristic is mapped on the Y-axis while the other is mapped on the X-axis. A curve is then plotted to illustrate subsystem performance relative to the two characteristics. Trade-off curves might be used to evaluate speed to fuel economy in the calibration of a given power train configuration, or the size of a radiator to its cooling capacity. In one particular case, an exhaust system supplier to Toyota created more than 40 different prototypes for one car program. Why make so many prototypes? The answer was to vary different factors and make tests to develop trade-off curves so that the Chief Engineer could understand the relationship of muffler back pressure to engine noise (see Figure 15-1) and make an informed decision.

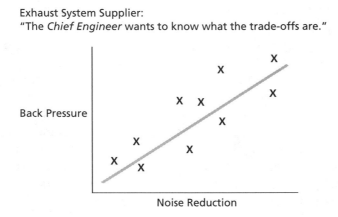

Figure 15-1. Multiple Prototypes Help Understand the Design Space

Trade-off curves are a fast and effective way to communicate about very complex and technical performance attributes in a set-based environment. An actual tradeoff curve used by the Chief Engineer of the first Lexus shows plotted on a graph of Max Speed (power) versus Gas Guzzler Tax where luxury cars from the United States, United Kingdom, and Germany sit compared to the stretch breakthrough target he set for the Lexus LS400 (see Figure 15-2).

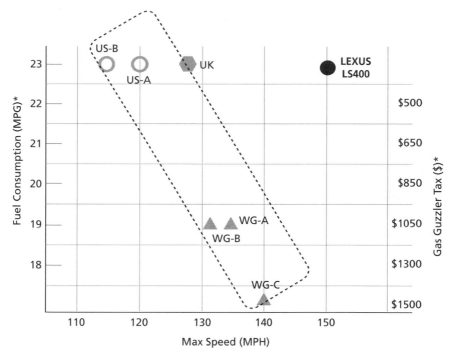

* Combined city and highway fuel consumption

Figure 15-2. Trade-off Curve to Set *Lexus* Breakthrough Target
for Speed and Fuel Tax

Decision Matrices

When engineers at Toyota want to consider various design alternatives or provide feedback or suggest solutions to design challenges, they communicate with matrices. As you will recall, one of the first lessons new engineers learn from their freshman project is to think problems through thoroughly, and part of this by process often includes presenting the team with a matrix. Once again this is a relatively simple but effective tool for communicating and evaluating alternatives. Design alternatives are listed on one axis of the matrix while specific evaluation criteria are listed on the other creating multiple cells. Each design alternative is then evaluated against those criteria and a quantitative or qualitative value is entered in the appropriate cell. For example an engineer might be evaluating design A, B and C against the criteria of cost, weight, durability, and functionality (see Figure 15-3). In this example, design versions A, B, and C are listed along the Y-axis while the evaluation criteria on the X-axis with corresponding values for each. Note that these values are not precise quantitative metrics but a subjective rating of achievement of the objective that allows for adding up the ratings across categories to a total score. This might not seem like much of an engineering tool, but it does give a rough ordering of the alternatives which is sometimes all that is practical . . . and it can be quite powerful for experienced engineers making decisions.

Design Alternative	Cost	Weight	Durability	Functional Performance	Total
A	1	2	2	2	7
B	3	2	3	1	9
C	2	3	1	3	9

Numbers 1–3 represent a simple rank ordering.

Figure 15-3. Sample Decision Matrix

Both of these tools not only make communication and evaluation more effective but provide a key tool for learning and preserving knowledge in the Toyota product development system.

Competitor Benchmarking Reports at NAC

Competitors provide an important and useful source of information in developing the concept for a new vehicle. NAC studies the popular or useful features and characteristics of competitor vehicles within the same vehicle class or market segment for vehicles it plans to develop. It benchmarks the competitor's vehicle sales and performance data against certain vehicle characteristics targeted for replication by the team. In some cases, the team uses vehicle teardown analysis to identify the underlying design enablers of those performance characteristics. NAC then summarizes the results of these investigations in multipage reports that program participants study. The team may follow up with a visit to the teardown facility and further examine those features or characteristics that fit within the vehicle concept; their findings are communicated to functional groups with recommendations for possible application to the vehicle that is being developed. The key phrase here is "communicated to"—information flow is clearly unidirectional and is not even flowing in the right direction.

Here again, NAC is copying a lean tool while missing the essence of the tool. The engineers who need to use these results do not "own" the teardown process. *Someone else in the company did the teardown and developed a report.* NAC has even subcontracted the competitive vehicle teardown process to outside firms. However, hired guns achieve only so much before moving on to other clients; even worse, they have no vested interest in providing data that is truly comprehensive and species specific to the companies with which they deal.

Because there is a cultural and professional separation between the people doing the analysis and the people who must synthesize and act on it, NAC engineers cannot fully benefit from these teardown exercises. This is a serious flaw in the traditional PD system. Even if engineers learn something and apply it to one vehicle program, what was learned and applied is rarely standardized or even communicated to other programs. As a result, learning tends to be isolated and disjointed. The opportunity to spread the learning organizationally has been lost.

Toyota Competitor Teardown and Analysis Sheets

Toyota uses a process that, on the surface, appears similar to NAC's. There are, however, significant differences in how that process is approached and applied. After the CE concept paper identifies high-level vehicle performance characteristics, the module development teams (MDT), discussed in Chapter 8, create component or subsystem level goals that are aligned with the system or vehicle-level performance goals. In a lean PD process, these component-level goals can be developed in several ways. One method is to use the team's preprogram plant visits to identify potential quality or manufacturability issues with the current component. Another method is competitor teardown analysis of best-in-class components.

Chapter 3 included a generic research example related to body development, and this example serves to illustrate how MDTs create and align component level goals. In the example, the module development team responsible for closures identified hem issues (the binding of the edge of the sheet metal on the door) on the previous version of its vehicle's doors. Reviewing assembly-plant quality data, the MDT found that the width of the gap or margin between the front of the door and the rear of the fender and the rear of the door and the front of the quarter panel varied to an unacceptable level. The group added this specific characteristic to its issues list. With benchmarking, the MDT was also able to identify features on a competitor's door with more consistent margins.

After taking margin measurements and compiling comparison data, they summarized their findings in a radar graph that identified specific areas of concern at both the rear and front edge of the door (see Figure 15-4).

The teams then methodically disassembled and analyzed the door assemblies for design or manufacturing characteristics that might lead to

Front Door Margin Comparison

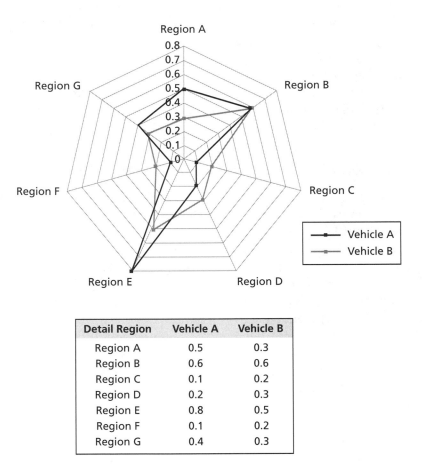

Detail Region	Vehicle A	Vehicle B
Region A	0.5	0.3
Region B	0.6	0.6
Region C	0.1	0.2
Region D	0.2	0.3
Region E	0.8	0.5
Region F	0.1	0.2
Region G	0.4	0.3

Figure 15-4. Radar Graph Comparison—Areas of Concern at the Rear and Front Edges of the Door

a better hemming condition. In this case, the MDT identified at least three characteristics that led to an improved margin and one that enabled superior flushness.

After completing these analyses, the MDT developed study drawings or *kentouzu* and a specific matrix plan for related, specific adjustments and modifications for the new vehicle. Styling, body engineering, and production engineering subsequently all signed this document and utilized it as a design guide.

There were two key success factors to the benchmarking: 1) it was owned by a team of engineers who specialized in and were responsible for the door, and 2) the benchmarking was turned into a specific problem definition with specific countermeasures. The same group that did the comparative analysis implemented the changes. *Ownership, responsibility, and good problem solving are all keys for a successful lean PD process.*

This scenario is repeated multiple times across the program as each group of functional engineering specialists on the program is responsible for its own benchmarking and tear down process. It is a hands-on process in which the engineers themselves literally crawl around competitor's vehicles busting knuckles and nicking fingers pulling off subsystems, reducing them to their component parts, mounting them and creating an in-depth technical analysis. The engineers enjoy the process and one of the side effects is that by taking a hands-on approach the engineers truly learn and usually a great deal more than just the narrow specification of a target part. Best of all, it is tacit knowledge, and it is *genchi genbutsu* at its best.

Standardization Tools at Toyota: Engineering Checklists, Quality Matrices, *Senzu*, Standardized Process Sheets

Standardization tools, such as checklists, are central to the Toyota PD process. Former NAC engineers who now work for Toyota are emphatic about this point, arguing that it is the main reason Toyota's PD system is consistently better than NAC's. The reason for this is that there is a process that makes these standardization tools add value. From the start, beginning with the *kentou* period, each functional activity utilizes a set of engineering checklists that guide decision making throughout the product development process. Checklists may define crucial steps within a process (process checklist) or provide guidelines for specific characteristics of a product design (product checklist). They are based on first hand experience and are updated and validated regularly to incorporate any new data or technological developments. In all cases, these checklists contain very detailed information about the product or process. Furthermore, the same groups that use the checklists maintain and update them at the end of each program and, as required, at *hansei* events.

As previously noted, one example of a product checklist is the quality matrix. The production engineering department, for example, maintains a specific quality matrix for each major sheet metal part on each vehicle (e.g., the *Camry* fender has its own quality matrix as does the *Sienna* minivan door outer). Listed across the top of the matrix are all the steps in the manufacturing process (see Figure 15-5). Listed down the left side of the matrix are quality and productivity issues potentially associated with this particular part, based on its vehicle functionality and manufacturing process. Complex parts may have hundreds or even thousands of individual considerations. By including quality as well as productivity issues, the team can begin considering "lean manufacturing issues" right from the start.

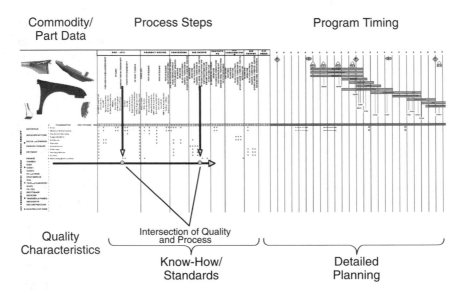

Figure 15-5. Quality Matrix and Know-How Database Integration

If a specific step in manufacturing potentially impacts a quality or productivity issue, a dot indicates this relationship. In the know-how database electronic version (V-comm) of the matrix, the user simply clicks on the dot and the appropriate standard(s) appear. In addition, each part's standardized individual development and test plan is available through V-comm. This electronic format gives easy access to products, processes, and cross-referencing that was previously unavailable.

The matrix contains specific recommendations for product design geometry as well as the limits of current manufacturing technology to produce a particular geometric characteristic. For example, the matrix might contain design parameters, such as the width-to-depth ratio to be maintained to stamp a certain feature. Two important characteristics of these standards should be highlighted: 1) they specify only those requirements that are critical to manufacturing, and 2) they usually provide standards in terms of ratios (depth to width) or "if-then" statements (if you require a certain flange angle and length, then this size radius will be required for quality bending). These characteristics provide solid guidance for effective manufacturing while supporting maximum design flexibility.

Senzu or the stamping engineering drawing is another example of a checklist/standardization tool. For each major stamped component, engineers have a three-view drawing. All the special manufacturing requirements are marked on the drawing near the relevant place on the specific part. (For example, the amount of springback compensation that must be built into the die to achieve the correct flange angle will be noted on the drawing near the point it is required. Or the amount of surface plussing and the blend pattern utilized to achieve an optimized surface condition will be noted on a class-one surface transition.) Like other checklists, these *senzu* are updated at the end of each program and utilized as a starting point for the next program. As described to us by one Toyota production engineer:

> "When we are close to the handoff to the plant we make notes on the *senzu* drawing. What was the anticipation versus what did we see? During the entire process, notes were kept by the process engineer and stamping engineer. Engineering changes are documented. Major documentation happens right before handoff to the plant. Mostly this is used for the next evolutionary cycle. If we discover something major we occasionally feed that back immediately but that is unusual. We put it in the file with the expectation that the next vehicle that is similar will use that information."

Another important standardization tool is the standardized process sheets that indicate the specific manufacturing process for each part.

(Figure 15-6 shows a standardized process sheet for a front fender.) By using these process sheets early in the PD process, engineers can design parts within standard manufacturing parameters. Exceptions are rare and typically require a *ringi-sho*, discussed in Chapter 14.

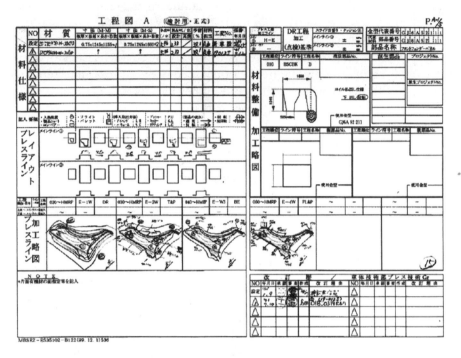

Figure 15-6. Standardized Manufacturing Process Sheets

Common construction sections are a standardization tool used to capture standard architecture for each part and provide design anchors for each vehicle. These sections enable the *kentouzu* (study drawings) generated during the *kentou* phase. They dramatically reduce the amount of work required of the body engineer as new design styles are considered.

The Role of Standardization and Learning Tools

Anyone familiar with lean thinking understands that organizational learning is one of Toyota's core competitive advantages. What is less understood is that organizational learning is only possible with living standards that are seriously followed and regularly updated. While competitors are working to become lean, Toyota's lean system continues to evolve simply because the company has built learning and evolution of standards into its systems. Each program, each component design, and each test conducted is an exercise in organizational learning through updating of lessons learned and standardized practices. Slowly and methodically making Toyota better, the power of organizational learning sometimes confounds attempts to measure Toyota's capability statistically in conventional measurement terms (development lead time, product cost, or product quality.) A better benchmark is the improvement trajectory from model change to model change.

Humans learn; technology does not. In a lean PD process, however, tools and technology can nurture and sustain human learning. But this can occur only if engineers with "towering technical competence" in specific functional areas take responsibility for using these tools to develop standards that become the company's new (and ever-changing) best-known way. It is up to leadership to empower employees to continually challenge and improve these standards until they become new standards. Whether these tools are manual or electronic, engineers responsible for a part of the vehicle must "own" as well as use these tools. It is this ownership that creates and cultivates a learning organization. Of course, without towering technical competence, a well-developed functional organization, and functional groups that take responsibility for maintaining and using standards, we end up with NAC's computer databases that are rarely used.

This chapter concludes the examination of the thirteen principles of LPDS and the authors' attempt to decompose the Toyota lean PD system into a LPDS model, a model that hopefully provides an understanding of lean PD's history, complexity, and power. Though the LPDS model on paper is no substitute for the application if LPDS in real-time, the purpose of this book is to explain the structure and lean tools of a strong and workable system. A key goal of this book is also to convey that while specific aspects of lean PD may be individually valuable, what makes them truly powerful is mutually supportive tools, processes, and human systems working in harmony. The next chapter describes how Toyota uses principles working in harmony to achieve a coherent system of product development; Chapters 17 and 18 continue this discussion and explore how your company can develop and implement its own coherent lean product development system.

LPDS Basics for Principle Thirteen

Use powerful tools for standardization and organizational learning

This chapter discusses some specific tools and methods that can help you leverage the power of tacit, "know how" knowledge and process, product and skill set standardization. These tools are not complex, and do not rely on expensive IT solutions. But they must be rigorous, clear, and owned, that is to say, maintained, validated, and updated by the functional experts who are expected to utilize them. Some of the tools discussed include the V-comm integrated Know-How database, engineering checklists, quality matrices, *senzu*, and standardized process sheets. The most crucial point about organizational learning and standardization is however, that the specific tools that an organization uses not nearly as important as the type of knowledge on which they are focused, the owner-ship of the learning process, and a strong cultural bias for learning and recognition of the true power of creative standardization.

Creating a Coherent Lean PD System

CHAPTER 16

A Coherent System: Putting the Pieces Together

"The way to build a complex system that works is to build it from very simple systems that work."

KEVIN KELLY

THUS FAR IN THIS BOOK WE HAVE ORGANIZED our discussion of Toyota product development into the three elements of the socio-technical model: people, process, and technology. We have used these theoretical concepts to simplify, codify and communicate what could otherwise be complex and amorphous. And while organizing our observations independently within these three categories aids in understanding Toyota and LPDS, it is not the way the real world works. People, processes, and technologies do not exist in isolation, insulated from each other and the outside world. In the real world Toyota's product development is an integrated evolving system, and the three subsystems of the LPDS model presented in this book interact, overlap, are interdependent, and work together to create a coherent whole. Changes to one subsystem will always have implications for the other two.

To illustrate this concept, take a simple mechanical system like an engine. It is quite possible to have the best piston, the best cylinders, and the best fuel injectors, but if they do not fit together, you have a bunch of great engine parts that accomplish nothing. Add the complexity of human systems into the mix, and you can see even greater implications. In product development the power of system "fit" is quite profound. In fact, we have found that the power of the system is often determined not by the effectiveness of any single subsystem, but the degree to which all three subsystems are aligned and mutually supportive. Moreover, product development efficacy is not determined by the competence of any single functional organization but by the seamless integration and collaboration of all of the specialists that reside along the product development value stream. This chapter examines both of these requirements: subsystem integration and cross-functional integration.

Many companies have strong individual subsystem attributes and highly capable functional departments. But all too often, subsystem characteristics do not fit together well, and functional organizations do not collaborate at the level required for high velocity lean product development. Often there is no "enterprise" perspective, the three subsystems grind against each other while individual functional organizations are insular and corporate cultures emphasize individual achievement driving incompatible objectives. They are in a state of constant conflict, as the organization seems to be locked in a perpetual struggle against itself. Consequently, efforts to implement lean, and lean product development processes in particular, go awry despite their best intentions.

We should emphasize that Toyota does not have complex system theory masterminds who orchestrate their very complex systems. In fact a hallmark of Toyota is the use of simple and common sense processes that work. Any of the subsystems that we have described through the thirteen principles is elegant in its simplicity. But taken together, the subsystems of Toyota's PD system are effectively aligned and mutually supportive to create a high degree of system harmony. Moreover, all of Toyota's functional organizations throughout the product development process are integrated and work collaboratively. The powerful synergistic effect produced by this is one of the most important reasons that Toyota's PD system is so successful.

Because the company consistently excels at integrating all of its parts, the whole runs smoothly. The three subsystems explored in this book—people, processes, and tools and technology—are elements of a larger, comprehensive and harmonious system with a single purpose. That purpose is to bring great products to market as quickly and efficiently as possible, something that Toyota manages to do consistently. Toyota has also learned that it takes the combined efforts of stylists, body engineers, manufacturing engineers, toolmakers, and manufacturing specialists to launch high quality products successfully. Integrating these diverse functional organizations and aligning their efforts is key to effective product development (Clark & Fujimoto, 1991; Wheelwright & Clark, 1992).

Thus, two things are necessary to create a high-performance lean PD system: 1) the integration of subsystems into a unified, coherent system whose single purpose is product development, and 2) the integration of the various groups of diverse technical specialists required to develop a new product. The greatest strength of a lean product development system lies in this seamless integration and in having a common culture that truly supports the lean PD system.

Subsystem Integration: People, Process, Tools and Technology

The first step in developing an aligned PD system is to establish what your customer values. Then you develop a waste-free workflow or *process* to deliver this value. However, efficient processes are useless if you do not have the right *people* available at the right time. You must consider the skills, practices, and organizational characteristics needed to execute the process that will deliver customer value. Finally, you must have the *tools and technologies* that support the activities of these people to enable them to achieve their potential and empower them to excel. In manufacturing and engineering, "cherry picking" single tools and practices without due consideration of the systemic environment is a common failing in many lean initiatives. However, as the opening quote of this chapter suggests, to build complex systems you must start with simple systems that work. As we will discuss in Chapter 17 you cannot attack your entire organization at once. You must take bite-size pieces and slowly transform work stream by work stream.

As discussed throughout this book, most consumer-driven companies are struggling with PD development systems that are out of sync and fall short of delivering customer value. Most companies do not invest in the integration of processes, people, and tools and technology, nor do they have a vision for a coherent PD system. One way to illustrate how purpose driven, aligned, and mutually supportive subsystems of a lean PD system succeed is to examine the integration of these subsystems from the perspective of the five lean principles of value, value stream, pull, flow, and perfection described in *Lean Thinking* (Womack and Jones, 1996).

Identifying Value: Delivering Customer-Defined Value

Delivering customer-defined value is the first and arguably the most important task in product development. If you do not create and deliver something the customer values, you have wasted your efforts. In lean thinking, delivering customer-defined quality becomes the core purpose of the organization. Toyota succeeds at this by focusing process, people, and tools on clear objectives. This process begins in the *kentou* or study period through the creation and diffusion of the chief engineer's (CE) concept paper. Based on the data and information gleaned from the CE paper, competitor analysis sheets, and preprogram plant visits, the module development teams set specific component level goals to achieve the CE concept. Various functional representatives support the CE's vision for delivering value to the customer; in addition, these individuals actively participate in the process by communicating the strategy to their respective home organizations. Furthermore, both engineering and manufacturing specialties are represented in the module development teams, increasing the odds the organization will actually deliver value to the customer.

These practices flourish within a culture that consistently practices a "customer first" philosophy. The flip side of this is that having a customer first context multiplies the effectiveness of your processes and tools. Clearly, people, process, and tools combine to support customer-defined value (see Table 16-1), and customer-defined value as a common culture supports people, process, and tools.

Table 16-1. Driving Customer-Defined Value

PROCESS	PEOPLE	TOOLS
• *Kentou* study phase to analyze and agree on a specific strategy to deliver value to the customer • Preprogram plant visits to identify any current quality issues	• Chief Engineer System • MDT cross-functional team to implement • Customer-first philosophy pervades the organization	• CE concept paper • Competitive tear-down analysis sheets for setting specific measurable goals

Enabling the Value Stream: Eliminating Waste and Variation

In *Lean Thinking*, Jim Womack and Dan Jones (1996) emphasize the importance of reducing or eliminating waste. Chapter 5 of this work elaborates on this theme, describing specific wastes that are particularly insidious to the product development process. Nonessential or redundant activities that add no value to the product are time and resource-consuming extravagances that become a serious drag on the value stream of any performance product development.

In a lean PD system, you need to eliminate waste found in and generated by your product development process. At Toyota, this begins with detailed scheduling and capacity planning, to identify the most efficient use of available resources, and with cross-functional agreement to a staggered data-release strategy to reduce the size of batches. Toyota's A3 standardized communication and problem-solving discipline provides for effective communication across functional organizations, minimizing much of the waste caused by poor communication.

Standardized best practices and their associated checklists drive consistent execution and support the broad responsibilities of the simultaneous engineer (SE). This, in turn, minimizes the need for hand-offs and builds in accountability. The up-front *kentou* or study phase allows problems to be identified and resolved before the execution phase. This practice, combined with design and process standards, minimizes engineering changes that are another tremendous source of waste. Toyota's checklists and standards, supported by culturally-ingrained process discipline, enable execution of detailed work, eliminating unnecessary reviews and quality inspections. Finally, as a learning organization, Toyota sees product development as a repeatable process that is subject to Plan, Do, Check, Act (PDCA). Every development program, and every stage within a program, is an opportunity to identify opportunities to eliminate waste in the next program.

The reason Toyota's PD system combats waste so effectively is that the various organizational subsystems reinforce each other. The SE could not function effectively were it not for the checklists, *senzu*, and standardized processes that minimize hand-offs. Nor could the SE be able to assume the great responsibilities required of him or her without first traveling a career path designed to develop appropriate skill sets for the SE role. The *kentou* process would not be nearly as effective if it were not for the participation of the cross-functional module development teams and tools like quality matrices and digital assembly. The point is that, in each of these cases, organizational subsystems work together to create a synergistic effect, which is also manifested in the product designs that drive lean manufacturing.

The *kentou* processes—early, intense involvement of the module development teams, supported by tools such as the part-based quality matrix, *senzus*, and standard manufacturing processes—drive quality and lean manufacturing to the very inception of the product. The combined power of all these individual components makes it possible for Toyota to exploit the full potential of lean manufacturing (see Table 16-2).

Table 16-2. Eliminating Waste in the PD Process

PROCESS	PEOPLE	TOOLS
• Minimize process hand-offs • Detailed scheduling and capacity planning for best resource utilization • *Kentou* to minimize changes and rework	• SE group and the career path that enables broad responsibility • MDT to anticipate and resolve issues early • Manufacturing personnel in up-front planning	• Checklists • *Sensu* • Standard process sheets • Quality matrices • Up-front use of digital assembly • A3 communication

Eliminate or Isolate Variation

In any process, task and arrival variation combined with system overuse are the greatest causes of long queues and subsequent protracted lead times (Hopp & Spearman, 1996). As one countermeasure, using best-practice templates in product development has a significant positive impact on throughput times (Adler et al., (1996). That is not to say that it is possible to eliminate all variation or that all variation is negative. On the contrary, some variation is inherent to product development, and some is even positive because it can lead to exploring divergent ideas. Recognizing these two peculiarities of variation led Toyota to the strategy of confining most variation to the preclay freeze phase of product development, thereby isolating it from the execution phase.

A PD system culture is shaped by the greater culture in which it exists. A culture that values adhering to standardized process and focuses on detailed execution drives a PD system culture that abhors delays. Toyota starts its PD process by convening module development teams far before clay freeze to identify and resolve core-engineering issues that are likely to cause delays in the execution phase. It uses schedules that minimize arrival variation, and Toyota engineers work to standardize processes and checklists, which provide best practice templates to reduce task variation.

Toyota treats tool manufacturing as a manufacturing operation and applies lean principles to the process, resulting in low levels of variation and WIP—high quality and short lead times. The value of team integrity has made it possible to automate the tool manufacturing process; Toyota toolmakers do not resist automation because of the trust created in the company's adherence to this value over the years. As technology advanced, the toolmakers became CAD modelers, process planners, programmers, and CNC machinists. These people, the process they are engaged in, and the tools they use combine forces to make possible low levels of variation, shorter lead times, and consistent, predictable outcomes for timing and quality (see Table 16-3).

Table 16-3. Reducing Variation in the PD Process

PROCESS	PEOPLE	TOOLS
• *Kentou* for core engineering and early problem solving before the execution phase • Standard processes • Lean tool manu-facturing process	• Cross-functional MDT gather before clay freeze • Disciplined to schedules • Trust from team integrity	• Checklists • Standardized process sheets

Flexible Capacity

As discussed in Chapter 5, the ability to create flexible capacity is critical to the highly cyclic environment of product development (Adler et al., 1996; Loch and Terwiesch, 1999). Stated simply, it is the ability to have resources available when required, without having to pay for them when they are not. This allows an organization to bring the right resources to the right place at the right time without the cost of negotiating with suppliers or the expense of staffing to peak workloads.

The primary enablers of flexible capacity and the use of capacity relief values at Toyota are *standardized skills, standard processes,* and *design standards.* This standardization makes it possible for technicians from pools to be assigned to a given program JIT, perform with little or no direction, and move on to another program as needed. (This standardization also allows Toyota to release work to affiliated companies with little or no transaction cost.)

Toyota's culture of discipline and the practice of rewarding individuals for process adherence, emphasis on team integrity, and rewarding technical excellence further reinforce flexible capacity. The company's cultural values promote the technical skills and support waste-free flexibility. If a company does not thoroughly understand and follow its standards, the practices that support flexibility will degrade into confusion, delay, and poor quality. Toyota extends these cultural values to its network of suppliers; companies that supply engineers to Toyota are treated as "partners" and the engineers are taught Toyota's specific engineering processes as well as Toyota's culture (see Table 16-4).

Table 16-4. Creating Flexible Capacity

PROCESS	PEOPLE	TOOLS
• Well-understood standard processes • Front-loaded capacity planning and capacity relief valves	• Career path designed to create standard skills • Affiliate companies are part of family • Technician pools • Reward people for process adherence	• Checklists • Design standards • Standard process sheets

Creating Pull and Flow

Once a company has reduced or eliminated waste and variation from a single process and streamlined the value stream, the next task is to make the remaining process steps flow (Womack & Jones, 1996; Rother & Shook, 1998). The goal here is to have a product move steadily from concept to customer without interruption or delay. In product development, this includes developing effective concurrent engineering capability along with synchronized cross-functional tasks.

Participants in Toyota's product development process stagger the release of product and manufacturing data so that data is pulled as required by the next functional organization while cross-functional teams maximize the utility of the available data; participants do not attempt work with unstable data. In the earliest phase of product development (*kentou*), for example, module development teams seek out data that, in some respects, seem related to things that may not be real-time concerns until months later. But the plant visits, matrices, and digital assemblies used by the MDT's *are* staggered, real-time activities. They are appropriate because they provide stable data that impacts the early stages of PD in a way that precludes problems in later stages or even in manufacturing. In addition, Toyota's cultural bias for detailed execution drives a thorough understanding of subtasks. This enables meticulous task design and synchronization, which, when combined with standards and standardization tools, eliminates review events. This creates a steady movement or flow of data from clay freeze to SOP. As discussed above, this is especially true in the tool-manufacturing phase of product development, when lean manufacturing concepts are utilized to create flow in Toyota's tool and die operations. The result is process flow and subsequent short lead times (see Table 16-5).

Table 16-5. Creating Flow

PROCESS	PEOPLE	TOOLS
• Staggered design release • Synchronized processes for simultaneous execution • *Kentou* early problem solving • Lean tool manufacture	• Detailed execution • Trust from team integrity • Broad knowledge of PD process • MDT	• Standard process • High precision machining and patented tools • Digital assembly early in the process

Enable Efficient Manufacturing

One of the three primary purposes of an effective product development system is to create designs and processes that support high-quality, waste-free manufacturing. Your greatest opportunity to impact the cost, quality, and manufacturing efficiency of a product is during its development. You access the true power of lean manufacturing when you apply lean principles to the product and process from the start. Throughout this book are examples and descriptions of ways in which Toyota's PD system acts as a key enabler for the company's production system. From the very beginning of the PD process, MDTs examine the impact of new designs on efficiency and ergonomics. Engineers utilize design standards, a modularity strategy, and standardized manufacturing processes to minimize variation and speed up the production ramp-up phase.

The staff production engineers and the plant production engineers belong to the same centralized organizations and share knowledge and goals. Production engineers keep their own checklists, which they share with product development at the beginning of any new program. In stamping, standardized tool and die manufacturing practices, like precision machining and construction, speed up setup times and minimize between setup variations (see Table 16-6).

Table 16-6. Enable the Toyota Production System

PROCESS	PEOPLE	TOOLS
• *Kentou* • *Hansei* • Precision die making to minimize setup and reduce manufacturing variability • Design around standard processes, common modular architecture (locators/construction sections/weld location) • Preprogram plant visits for designed-in countermeasures	• Cross-functional MDT activity w/production pilot team leader • Production engineering a single, centralized organization. Plant and engineering staff have common goals and experience • Manufacturing is a priority	• Quality matrices • Digital assembly • Patented machine tools • Quality matrices with quality and productivity issues • Digital assembly tools at *kentou*

Perfection: Build in Learning and Continuous Improvement

"Pursue perfection" is the final step in lean transformation advocated by Womack and Jones (1996). It is an interesting expression, implying that there is no end to how much better a process can become through the consistent application of lean methodologies. Lean is a journey, not a destination, and only by learning and continually improving can organizations compete in the hypercompetitive product development environment that currently exists. It bears repeating that learning is a fundamental part of Toyota's DNA (Spear & Bowen, 1999) and that Toyota's learning is about acquiring, sustaining, and passing on valuable tacit knowledge. In a lean PD system, numerous tools and technologies, processes, and organizational practices support the culture characteristics that create a powerful learning network. This book has covered a few of these, showing how each of them has been consistently communicated, applied, and practiced at Toyota (see Table 16-7).

Table 16-7. Building in Learning and Continuous Improvement

PROCESS	PEOPLE	TOOLS
• In-process reflection/*hansei* events • OJT and mentoring	• Technically-skilled managers • Learning DNA • Skills based career path and evaluation • Teaching as leadership • Users update standard checklists	• A3 problem solving • Learning focus • Supplier technology demonstrations • Know-how database • Freshman project • PDCA cycle

Cross-Functional Integration

Aligned and mutually supportive lean product development subsystems are a solid foundation for a coherent PD system that is more powerful than the sum of its parts. However, product development performance also depends on the integration of functional organizations without diluting the specific assets and good characteristics of individuals in the functional organization being integrated. Bringing high quality products to market successfully requires the efforts of a large group of diverse technical specialists. There must be appealing styles from the design studios, practical and workable designs from product engineering, capable processes from manufacturing engineering, quality tools on time from tool and die, and high quality products from manufacturing. You create great products when all of these functional activities are goal aligned, synchronized, integrated, and reinforce each other. The degree to which they act harmoniously toward a single goal determines the success of the enterprise. Seven cross-functional integrators that have evolved at Toyota were discussed in previous chapters (particularly Chapter 8). They are worth revisiting in connection with the information presented in this chapter.

Integrated Development Teams. Toyota's cross-functional development was formalized fairly recently, but similar teams have existed for decades, working informally to meet common goals. The teams consist of representatives of functional specialties involved in the early development of product and processes when decisions have the greatest impact on the ultimate success of the program. Because the participants remain a permanent part of their respective functional organizations, the danger of lost or unshared technical knowledge across programs is minimal.

Obeya *Team System.* Another recent development at Toyota, the *obeya* team integrates various product development participants throughout the life of a program. Managers of each functional organization meet in a war room with the CE and staff several times a week; these meetings enable fast decision making and information sharing across functions.

Checklists and Standardization. This is a somewhat surprising source of integration at Toyota because it seems to reflect what occurs during a single process or operation. But when functional groups work reliably to a standard process or design standards, other groups know what to expect from them and when to expect it.

Functional Build. Previously discussed in Chapter 13, functional build is a key cross-functional integrator because it brings together various functional specialists to make tactical decisions based on system-level criteria. Diverse technical specialists work in close physical proximity to the vehicle and to each other, sharing technical communication across functions. Each specialist contributes knowledge from his or her respective organization, but the group makes decisions based on objective data and a single purpose.

A3 Standardized Communication. Accurate communication is crucial for integration. Given the number of specialized groups in a product development project with its own jargon and terminology, this can be difficult to achieve. Because the A3 format uses graphics and short, precise phrases, it bridges these differences and, quite literally, puts everyone on the same page.

Resident Engineers. Toyota temporarily assigns resident engineers from suppliers both within and outside of Toyota for periods of up to three years. These temporary assignments, especially to affiliated companies, have a positive effect on integrating the various organizations and spreading standardized operating procedures and technical knowledge. This practice promotes a sense of belonging to a common corporate entity.

Hansei *Events.* During these reflection events, different groups come together to discuss performance on an ongoing or recently completed program. This event not only provides an opportunity to learn and improve, but an opportunity for cross-functional groups to identify opportunities for better coordination of their respective activities. Discussing common outcomes helps different functions build a sense of shared destiny.

Toyota does not leave cross-functional integration in product development to chance. It uses these seven mechanisms to integrate the many diffuse technical specialists required to develop a new vehicle, without the risks posed by the platform team approach. Through these integrators, its product development subsystems, and its culture, Toyota has evolved an integrated, aligned, and coherent PD system. The operative word is "coherent."

CHAPTER 17

Eliminating Waste in the Product Development Value Stream

*"I clearly remember an argument with my father about starting a
new project. I was against the project, and I won the argument. But
then my father asked me to give the project a try anyway, which I did.
Much to my surprise, it turned out to be a great success. From then
on I did less arguing and more practicing."*

KIICHIRO TOYODA (referring to his father Sakichi Toyoda)

THE PREVIOUS CHAPTERS OF THIS WORK presented the principles and framework of the lean product development system, supplemented by illustrative examples of specific methods and cultural practices that enable or limit the potential of this high-power strategy. At this point, the reader may well be wondering how to use the information in this book to transform your existing PD system into a lean PD system. In the next two chapters we will provide tools, methods and insights designed to aid you in this challenging and exciting journey. This chapter focuses on improving the product development process through a value stream focus and the application of lean PD process principles. Chapter 18 will examine proven methods for transforming to a lean culture.

As posited by Rother and Shook (1998), "A value stream is all the actions (both value added and non-value added) currently required to bring a product from raw material to the arms of the customer or through the design flow from concept to launch." Value stream mapping (VSM) is an effective technique for graphically drawing these activities, as well as the flow of information and product between those activities. (Toyota, which developed this concept for manufacturing, calls it the Material and Information Flow Diagram.) By mapping the value stream, you can visualize the entire process, which then supports fundamental process reinvention that requires you go forward to the most critical step—developing a future state value stream map. The current state is the foundation. The future state is the vision of where you want to go. For Rother and Shook, VSM is a proven tool for improving manufacturing processes and the missing ingredient in many failed lean manufacturing initiatives. The same authors explain that VSM is powerful because:

- It helps you visualize more than a single process.
- It helps you see more than waste—it helps you see the sources of waste.
- It serves as a common language for all participants.
- It brings you beyond fixing individual problems in the future state to a coherent vision of a future state system.
- It forms the basis of an implementation plan. It helps you design the whole system and becomes a blueprint for lean implementation.
- It makes decisions about flow apparent.
- It shows the link between information and material flow.

Central to value stream mapping is a systems philosophy of managing change. Since you are seeking to change a complex system, it is not enough to identify isolated problems in particular processes and fix those problems. Therefore, value stream mapping never stops with a current state map. *The current state map provides grounding in reality, but the real leap is in developing a vision for the future state.* This is what provides a vision for the future system and the subsequent action plans you must implement to achieve it.

Although Rother and Shook include product development as a part of their definition of VSM, one of the authors of the current work (Morgan, 2000) was the first to develop a specific adaptation of VSM for the product development environment (PDVSM). While focused on product development, the methodology can be applied to any process that crosses functional or organizational boundaries and where tasks are parallel and interdependent. Because it is so critical to the change process, an overview of the PDVSM tool merits further attention, both in the context of this chapter and in a case study in the appendix demonstrating its use. A separate PDVSM workbook (also by Morgan) provides a step-by-step explanation of the PDVSM process.

Product Development Value Stream Mapping (PDVSM)

Many of the issues endemic to complex processes are particularly problematic in the PD process, and the PDVSM tool in product development is at least as useful as (or perhaps even more useful than) VSM is in manufacturing. Here are some reasons why this tool is so powerful:

1. *Task and arrival variability resulting in long queues and wasteful work and data-in-process inventories are pervasive in the PD process.* Although some variability may be inevitable, even beneficial, due to the nature of design work, the authors' studies and the previously-mentioned research by Adler demonstrate that it can be managed.

2. *Non-value added activities or wastes are rampant in the PD process just as they are in traditional manufacturing processes.* The longer time frames and complex nature of the PD process work tends to obscure a great deal of insidious non-value-added activity.

3. *Discernable pattern of product evolution from one state to another over time.* However fitful, the PD process does progress from concept to customer. In fact, the PD process consists of many progressive flows of information and decisions and parallels similar issues in manufacturing (e.g., batching versus single piece flow).

4. *Capacity and scheduling-related issues.* System utilization is one of the best predictors of lead times in manufacturing and product development systems. Whether measured in person-hours or throughput, both types of processes must deal with capacity constraints. Product development systems typically have big peaks and valleys in workload, and the peaks operate at levels far beyond their capacity.

5. *Hand-offs from one functional activity to another.* The greatest challenges are often found at the intersections of activities, whether in manufacturing or PD.

6. *There is a work process that you must continually analyze, standardize, and improve.* Although the nature of product development work and manufacturing work differs, both can be enhanced through process improvement efforts.

7. *Pressures for lead-time reduction.* The focus of value stream mapping is on improving cycle times and time in system, especially as compared to the actual value-added time. Continually achieving shorter time to market is a system level goal of high-performance product development systems.

8. *Tasks must be synchronized.* In product development, you must synchronize concurrent tasks across functional organizations or work centers to minimize the waste of rework and maximize the benefits of concurrent or simultaneous engineering.

9. *Constraints must be identified and managed.* PD processes, like manufacturing processes, are only as good as the weakest link.
10. *Creating flow.* Once you have eliminated waste, you must make the overall process flow by synchronizing cross-functional tasks and identifying your constraints. This is as important in PD as it is in manufacturing.

Just as with manufacturing processes, the PD process can be improved by value stream mapping and diligent implementation of the resulting action plan by a cross-functional team.

Addressing Some Differences Between PD and Manufacturing VSM

While manufacturing and PD processes may have some common characteristics, product development and engineering environments are significantly different than manufacturing, and we have yet to find a company that has been successful in improving PD performance simply by moving shopfloor tools to PD. To use PDVSM effectively you must understand some of the unique aspects of the PD environment and how it contrasts with a manufacturing environment.

1. *Product development is largely concerned with the flow of data rather than a physical entity. The main focus is on the data value stream and data are more elusive than physical raw materials.* Later in the process we look at tool and die making and other manufacturing processes but early on it is all data. An important characteristic of data is that it can reside simultaneously in multiple locations, allowing many of the activities in PD to operate concurrently rather than in sequence.
2. *Temporal measures in PD are calculated in weeks, months, and even years (as opposed to minutes and seconds in manufacturing); are often ill defined; and can be highly variable.* This has important implications for both the mapping process and the learning/continuous improvement process.
3. *The nature of the work in PD is intangible compared to that of traditional manufacturing.* Much of the work is intangible, even invisible. It is what Peter Drucker referred to as "knowledge work." By its nature, it is more diverse and less predictable.

4. *Data, information, and product flow is often nonlinear and multidi-rectional.* Much of the work is iterative, cyclic, or narrowing in nature. Rather than a continuous, consistent march from raw material to finished product, the PD process is a fitful progression, punctuated by significant setbacks. Information and data flow reciprocally between multiple activities simultaneously in the PD process. The PD process is not as predictable as traditional manufacturing and requires substantially more and different types of communication.

5. *There is a larger, more diverse group of participants in the PD process.* The PD process requires many technical disciplines or functional activities, each of which contains multiple tasks and subtasks.

Although these issues differentiate the typical PD processes from manufacturing processes, be aware that the PD process inherently shares many of the characteristics that occur in all multistage processes.

Specific Challenges and Countermeasures for Mapping the PD Process

Table 17-1 shows some of the unique characteristics of the PD process that present significant challenges for a mapping methodology. You will need to address these challenges individually, along with specific mapping techniques as countermeasures.

Table 17-1. VSM in Product Development Versus Manufacturing

Product Development Process	Traditional Manufacturing Process
Virtual data flow	Physical product flow
Weeks and months	Seconds and minutes
Primarily knowledge work	Physical manufacturing
Nonlinear and multidirectional flows	Linear evolution
Large, very diverse group of technical specialists	Primarily manufacturing organization

Virtual data

In a manufacturing environment value stream mapping focuses primarily on the material flow. Information/data primarily supports the material evolution from raw materials to finished goods. The tendency of physical materials to reside in one place at a time definitely simplifies the mapping process. In product development, you need to be concerned with the evolution and flow of ideas and data. The fact that data can reside simultaneously in multiple locations is crucial to concurrent engineering (see Figure 17-1) but makes the mapping process (as well as coordination of activities) somewhat more complicated because there are simultaneous activities that you must display on the map.

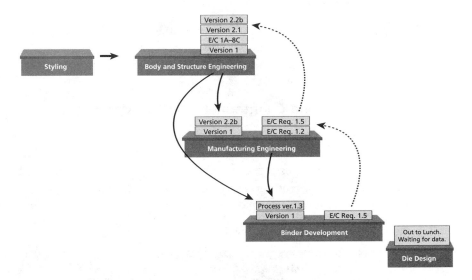

Figure 17-1. Virtual Data Resides Multiple Places Simultaneously

You can address this difficulty by creating multiple-functional tiers or horizontal layers, aligned with a single time scale, in which you can see many different activities occurring simultaneously at any given time during the PD process (see Figure 17-2). This will give you system level insights into the process. From this perspective, VSM is useful for facilitating concurrent engineering.

Simultaneous activities also complicate your ability to "go and see" or "go to the source," as is recommended in traditional value stream map-

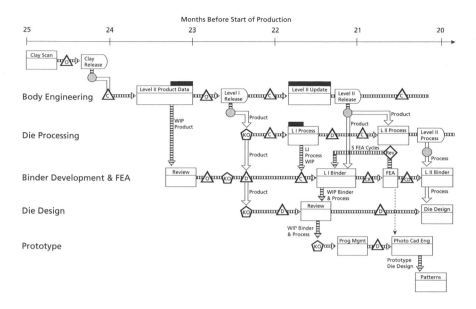

Figure 17-2. Time Scale and Functional Process Layering Applied
to PD Value Stream

ping. After all, people, like materials, cannot reside in more than one place at a time. However, the longer task time frames associated with the PD process gives you sufficient time to record multiple functional activities through engineering activity logs and interviews as well as to employ various database technologies to capture movement of virtual data instantaneously.

Longer time frames

The longer, more variable time periods associated with the PD process is another important distinction from typical manufacturing where task times are measured in minutes or even seconds and are usually stable, predicable, and subject to little variation. You measure the task times in product development in weeks, months, and even years, and these can be quite variable (see Figure 17-3).

For this reason, PDVSM tacks the time scale located at the top of the page on Figure 17-4 so that you can draw up all activities with respect to this scale. It is also possible to locate major scheduled milestone events to which you can compare actual events (see Figure 17-4).

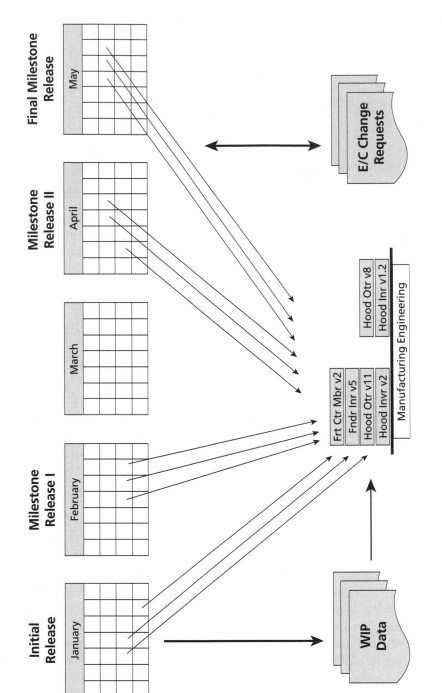

Figure 17-3. Product Data Release Time Frames

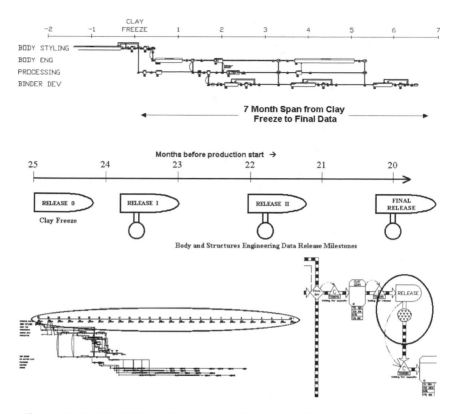

Figure 17-4. PDVSM Data Release and Program Milestone Representation

As the level of mapping focus changes, you should change the time scale accordingly. When you are mapping higher-level cross-functional activities, for instance, a time scale in weeks or even months is appropriate. If you zoom in on an activity to map a specific process, your time scale will probably be in working days (see Figure 17-5).

Knowledge work

In traditional manufacturing, work takes place on the factory floor in the form of employee labor and machine processing. In contrast, much of the work involved in product development is knowledge work, occurring in the worker's mind. If you are not documenting this work, it will be invisible and therefore difficult to map. You can collect timing data for your current state map for slow moving physical objects such as dies or fixtures using tool tags. Figure 17-6 is an example of a tool tag that was used to

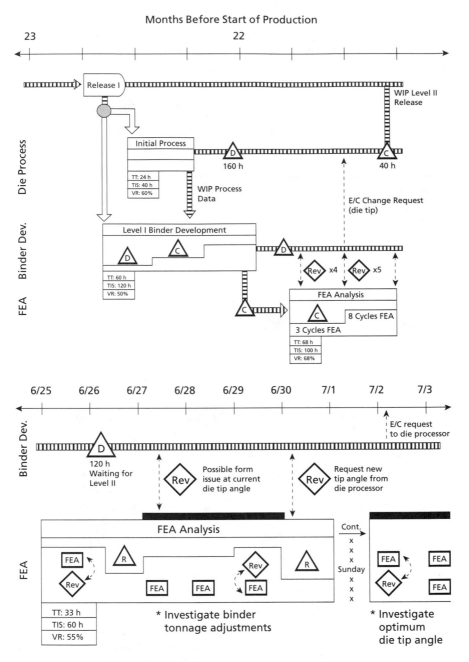

Figure 17-5. PDVSM Different Levels of Magnification
(month, upper, versus days, lower)

Tags are affixed to the tool and updated by plant associates or material handlers every time an activity is performed on the tagged tool.

Front of Tag

Manufacturing Activity Log				
Part # _____				
Date	Start Time	End Time	Operation Category	Operation Description / Comments
			☐ Setup/Teardown ☐ Machine Cycle ☐ Manual Activity ☐ Re-setup ☐ **Delay/Queue**	Track for activity (value added) and track for duration of setup, machine and manual processing (and any delays).
			☐ Setup/Teardown ☐ Machine Cycle ☐ Manual Activity ☐ Re-setup ☐ **Delay/Queue**	
			☐ Setup/Teardown ☐ Machine Cycle ☐ Manual Activity ☐ Re-setup ☐ **Delay/Queue**	
Date Verified: _____ Verified by: _____				

Back of Tag

Tooling Movement Log				
Date	Time	From	To	Move Description
				Track for movement in plant.
				Log TO and FROM data, date and time, and reason for the move.
Date Verified: _____ Verified by: _____				

Figure 17-6. Example Data Collection Using Tool Activity Tags

follow the flow of tooling development over time and generate activity logs with actual times. You can record data on timing in engineering activity logs, shown in Figure 17-7, for binder development. Figure 17-8 shows the actual hand-drawn current state map of the binder development process from this data.

Although logs and interviews have their limitations, you can support them through cross-referencing and the data-tracking technologies discussed above. Virtual and physical manifestations of the knowledge work are also available as evidence of work performed. Virtual manifestations include computer graphics of part designs, test result animation, and data. Physical manifestations (such as clay models, drawings, prototype parts, tools, and finally the finished product) also serve as indicators of project status, location, or quality.

We should note that for purposes of improving the process extremely precise data is not always necessary. Ballpark estimates of times can be a basis for seeing large amounts of waste and a future state vision can be constructed to significantly improve the process. Often very effective value stream mapping workshops are conducted with cross-functional teams and estimates by participants are used to supplement existing data.

Complex information flow

In manufacturing, the value stream is typically linear and unidirectional, marked by discrete transition from one state to the next until the finished good emerges from the process. A specific process step, once completed correctly, does not have to be repeated. In product development, however, much of the work is iterative, with data and information flowing back and forth between functions in a complex web of activity. When combined with the many different types of data and information flows, this can add more complexity to your mapping assignment.

Data flow that is important to capture in product development includes primary data releases, feedback, engineering changes, scheduling information, and "unofficial" data exchanges. You need to identify decision points and process iteration because both cause major queues and serious process delays. There are many different types of data/information flows to map:

- The product data itself while it is in a virtual state. This includes both partial and complete product data release dates as well as engineering changes.

Month	Week	PART STATUS SUMMARY	ENGINEERING ACTIVITY	Act/Tot (Hrs)
1	1	Clay freeze		
	2			
	3			
	4	WIP data released by body eng. Waiting for official work order to start	Waiting for work order	0/40
2	1	Scheduled milestone I missed by body eng	Waiting for work order	0/40
	2	Milestone I data released. Got OK to review WIP data despite no work order	Waiting for work order. Review initial WIP data	8/40
	3		Waiting for work order	0/40
	4		Waiting for work order	8/40
3	1	Milestone II missed by body eng. Level I process received from die processing	Waiting for work order	0/40
	2	Work order received. Immediate kickoff. Pulled most recent version of product from body eng	Update binder development to level I status	32/40
	3	Milestone III release. Pulled level II data from body eng. Possible formability issue at FEA—body eng notified	Waiting for FEA reviews. Update binder development to level I status. 3 versions of binder submitted to FEA—three cycles performed	16/48
	4	Formability issues at FEA. Discussing with body eng on course of action. Reworking binder to try to resolve issue. Working with FEA	Waiting for FEA reviews. 4 different versions of binder submitted to FEA—6 cycles performed	20/40
4	1	Milestone III release missed by body eng. Reworking binder to try to resolve formability issue. Working with FEA	Waiting for FEA reviews. 2 more versions of binder submitted to FEA—2 cycles performed	20/40
	2	Waiting for updated data	FEA ran 2 more cycles—no luck	0/40
	3	Updated level II data received from body eng	Rework level II binder to account for updated product. Ship updated binder and die design for their kick-off meeting	40/43
	4	Milestone III release	Waiting for data and FEA review. Prepped and shipped new binder to FEA to verify formability—1 cycle performed	8/40
5	1	Pulled final data from body eng	Working on other parts. Update binder to final release	32/40
	2		Working on other parts. Update binder to final release. Shipped final binder to die design + FEA for final check. (FEA is busy with other parts.)	16/40
	3		Waiting for binder development review meeting with customer. FEA ran 1 cycle with new binder—no issues	0/40
	4	Stamping plant has issued change in manufacturing line. Binder development is unaffected	Review meeting. No issues.	0/40
6	1			0/40
	2			0/40
	3			0/40
	4			0/40

Figure 17-7. Example Activity Log Summary, Binder Development Activity—Door Interior Panel

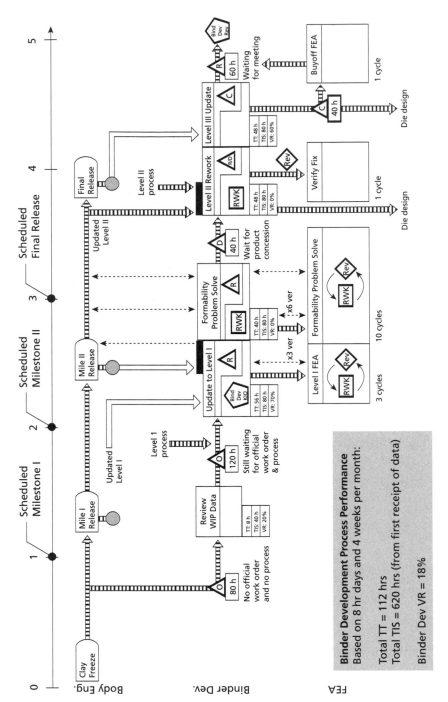

Figure 17-8. Binder Development Current State VSM

- Administrative direction or information given by control organizations, such as the program team. This might include milestone dates and status to schedule, engineering change authorization, decisions and approvals, purchase orders and quantities, bills of material, etc.
- Feedback information communicated in response to some program development activities. Examples might include manufacturing feasibility feedback on a part design, or feedback from some functional integration event, such as prototyping or report out events on status to schedule, and so on.

You must document each of these information flows in the mapping process, differentiating them by color, type of lines, and specific icons. You must simultaneously engage in the process of narrowing the data. Narrowing refers to the process of reducing a number of potential design solutions until only a single selection remains. This process also takes place over time and you can illustrate it on the map using a single color to show multiple activities or solutions working simultaneously within the same function. These activities are reduced over time until only one solution remains and is then passed on to the next activity.

Large, diverse group of specialists

The PD process requires the efforts of large and diverse groups of technical specialists (see Figure 17-9). Pure manufacturing usually involves only manufacturing people, either in direct labor or in a support role.

This difference clearly illustrates why PDVSM is so critical to "seeing" your process. It will assist the many functional organizations to identify the complex web of product development interdependencies, and, as previously mentioned, enable effective concurrent engineering by identifying specific activities across functional organizations at any given point in time.

PDVSM Workshops

Utilizing engineering activity logs to obtain the data required is an accurate way to create a current state PDVSM. However, because product development cycle times can be fairly long, this can be time consuming and should be used only if accuracy is critical. An alternative way to create both the current and future state maps is in PDVSM workshops. These three-day workshops were developed by Morgan (2001) as he began to

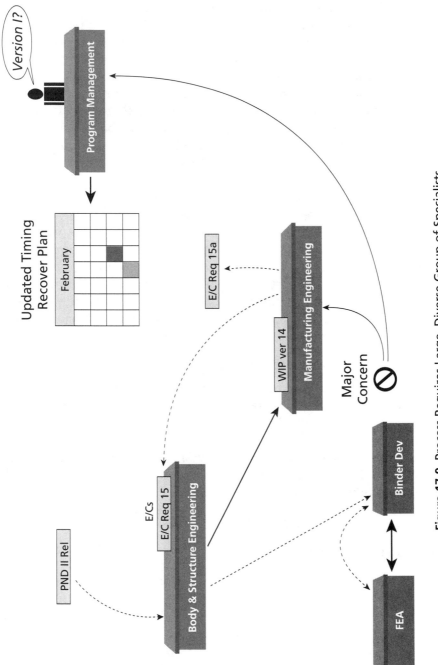

Figure 17-9. Process Requires Large, Diverse Group of Specialists

work with companies on improving existing PD systems and have proven to be effective in a variety of product development environments (see Figure 17-10). Gathering data through tags and logs can lead to highly accurate data, but the process can take weeks or even months. PDVSM workshops require only some preliminary work and using estimates of times can be completed in a few days. Figure 17-10 shows a sample agenda from an actual three-day PDVSM workshop. One advantage of these workshops is cross-functional dialogue that activity logs do not permit. Getting a cross-functional team to engage in deep dialogue focused on a common process, developing common objectives for performance, can be quite powerful in and of itself. (Of course, having collected data via activity logs a workshop format can still be used for constructing the value stream maps.)

The first step in organizing a workshop is to choose a small knowledgeable group to identify what product or product family will be mapped, the level of detail required, and the start and end points. This is *scoping the project.* Once the group understands this, it will need to identify the organization's customer and what value the company delivers to that customer from the process. Based on this information, the group can determine who should attend the workshop. Participants from all process functions that you are intending to map in the workshop should be present or represented, including the internal or external customer of the process. The needs of the customer and the results of the process should be aligned. Therefore, before having your participants start the workshop, you need to gather supporting data from the target mapping project and create a map framework (similar to that discussed for PDVSM) from activity logs. Once you do this prework, the three-day workshop will go something like this:

- *Day one PDVSM workshop.* The first day of the workshop begins with a brief discussion of PDVSM and the scope and objectives of the workshop. The participating team lays out the project and reviews the map framework that was created during prework. All functions are laid out along the Y (vertical) axis of the map framework and the correct time frame is presented along the X (horizontal) axis. (The authors typically print this framework on paper that is 3 or 4 feet wide by 5 or 6 feet long.) Then the team begins to recreate the process, utilizing large "sticky notes" for process boxes and drawing by hand the information flows. The team should have

PDVSM Workshop Framework

PREWORK	WORKSHOP			
SCOPING	**DAY 1**	**DAY 2**	**DAY 3**	**Management Report Out & Commitment**
Overview of Lean PD	Overview of VSM "A tool for improving processes"	Lean Product Development Principles Discussion	Time Bounded Implementation Plan	72-hour response
Who is the customer?				
How does the process deliver value?	Data for Current State Map	Current State Map Opportunities/ Identify Waste	Implementation-Enabler Proposals	**Follow-up**
What value is created?				
Workshop goals				Status update to management on 30 day, 60 day, 90 day, 1 year
What to map	Workshop Current State Map Creation	Future State Map Creation		
Start and end points				
Who should participate?				
What data is required/ available?				

Figure 17-10. PDVSM Workshop Framework Agenda

brought with them all relevant historical data so they can draw upon it to maximize the accuracy of the map. You can employ a number of tools, techniques, and methods to increase the chances of a successful mapping event, but that is beyond the scope of this chapter. If you scoped your project correctly and the team's first day of the workshop was effective, they should have a completed current state map and a very tired but excited group of people.

- *Day two PDVSM workshop.* On the second day, the team begins to create the future state map. There should be a great deal of energy and anticipation from the first day of the workshop where participants began to see some of the opportunities for improving their PD process. This is a natural result of "seeing" that process for the first time. The team needs to resist the temptation to start improving the individual pieces of process immediately. The authors forestall this temptation by a discussion session focused on the 13 principles of lean product development covered in this book, giving particular attention to the principles, methods, and techniques addressed in Chapters 3 to 6 (the process section of this book):
  - Create leveled flow and eliminate rework in your process by front-loading problems.
  - Design quality into the product design, achieving system compatibility before component completion.
  - Create cadence mechanisms to drive the process forward together.
  - Synchronize cross-functional activities to transfer the right information at the right time.
  - Integrate your suppliers into the process.

  This discussion will lead the team to identify additional opportunities for improvement and provide proven countermeasures to address those opportunities. As they study the current state map, participants should be actively engaged in questioning what they are looking at, asking "where are the wastes?" and "how can you use the principles to develop effective countermeasures?" Once this is completed, the team can begin to create the future state map of their process.

- *Day three PDVSM workshop.* On day three, team members work on a time bound implementation plan based on transforming their process to the future state. They must identify specific process enablers that will allow or "enable" the improved performance level of the future state process. How, specifically, will you do work differently to make the future state a reality? Proposal A3s are an excellent tool for this purpose. You combine the individual A3s into a detailed implementation plan that includes metrics and then develop a PDCA approach to executing the plan. Those A3s are some of the same documents that are posted in the *obeya* and rigorously debated over at the cross-functional design reviews.

Finally, the team reports to senior management, those empowered to make necessary changes or purchase required tools or technology. The company should set a time limit (within 72 hours) for the senior management team that has been involved from the beginning of the project to respond to the team's proposal.

Learning to See Product Development as a Process

The PDVSM process may be the first time that the participants are able to truly *see* their PD process, and, in that sense, it is an invaluable learning tool. The knowledge produced by the PDVSM exercise alone is well worth the time required. Moreover, PDVSM is a communication tool, serving as a common language between disparate functional organizations. For the first time, functional activities can begin to understand their respective roles within the whole system context and the effect that their actions have on other functional activities. The workshop format brings these functional organizations together and focuses them on the process—not on fault finding. A common language facilitates a deep understanding of your product development process. Without this, system level, holistic improvements are impossible. PDVSM also brings out cause and effect relationships, making them more evident. The long delay that normally exists between cause and effect in the PD process disappears when reduced to a value stream map. Lastly, PDVSM enables you to *see* overlapping concurrent PD activities, showing all on-going activities in the PD process at any given time. This helps you to gain significant insights into effective management of concurrent activities as well as opportunities to create greater concurrency. PDVSM is perhaps the most effective tool for improving PD concurrency because of its ability to display multiple functional activities simultaneously at different levels of detail.

It is important to remember that learning and continuous improvement cannot stop with PDVSM and your process redesign efforts. One of the hallmarks of a lean system is continuous learning and improvement. It is critical that you think of this improvement effort as the beginning of a completely new way of developing products and not as an end in itself.

Changing the PD process is necessary but not sufficient to create a lean product development system. In fact, changing the process may be the easy part. In the next chapter we will examine what it takes to make real change in your culture and people systems.

CHAPTER 18

Getting to Culture Change: The Heart of Lean PD

"Certainly, the thieves may be able to follow the blueprints and produce a loom. But we are modifying and improving our looms every day. By the time the thieves have produced a machine from the plans they stole, we will have already advanced well beyond it."

KIICHIRO TOYODA responding to theft of blueprints from

Toyoda Automatic Loom Works

ONE OF THE BIG MISTAKES many companies made in applying TPS to transform traditional manufacturing environments was to view TPS as a tool kit. Value stream mapping described in Chapter 17 is a powerful tool for understanding the process flow and where to reduce waste in the value stream, but it is just a tool. And it focuses mainly on only one of the three subsystems of the LPDS model—the process subsystem. Unfortunately, if the transformation process ends with a few workshops and a few improved value streams, your efforts will be mostly for naught. Without real culture change your value streams will start to fill back up with waste and before long look much like where you started. Many companies ask us how to "sustain the lean changes made" and that is the wrong question. The issue is not putting in a technical lean fix and then having some magic chant that makes the change stay in place. The organizational culture needs to be changed so that making improvements and having the discipline to follow the best known procedure is a way of life.

In recent years, the authors have worked with a number of organizations interested in developing a lean product development system modeled after the Toyota PD system. Most of these organizations are just a few years into the journey. All have discovered that the transformation requires some radical changes or *kaikaku* within existing PD systems. While working with these organizations, the authors have also learned a few things, including some basic truths of lean PD transformation:

1. *Transforming PD into a lean process is more complex and less precise than transforming manufacturing into a lean process.* There are so

many variables, parallel activities, interdependent paths, and complex feedback loops that it is impossible to model the product development process precisely. Value stream mapping, for example, can be used both in lean product development as well as in lean manufacturing, but it does not yield the same precise results. Changes in cycle times, up times, takt times, etc., are never as precise in product development. This does not mean it is a futile task; it is simply a different task than changing a manufacturing process requiring a different approach. Although there are limits to the degree of precision, dramatic improvements are possible.

2. *Cultural issues increase the complexity.* Holding workshops and developing value stream maps with detailed plans of action may be complex tasks, but they are relative child's play compared to dealing with all of the cultural issues that must be resolved before a company can create an environment that supports lean product development.

3. *Engineers are engineers and tend to want to reduce lean PD methodologies to technical tools.* This does not work. Transforming to a viable lean product development system requires much more than a set of sophisticated tools; it necessitates a human system renaissance.

4. *Senior leaders must be intensely involved in the transformation process but they typically do not engage at a significant level.* The importance of senior leader commitment to lean manufacturing transformations has been well documented, but getting this commitment is a challenge in and of itself, even more so than in efforts to transform manufacturing. Because implementing change to a lean PD system is a journey filled with complexities, high levels of risk, and far-reaching organizational implications, the committed participation of senior leaders is essential.

5. *Senior leaders must understand the commitment and have patience.* Patience is a luxury few leaders seem to be able to afford. If lean PD was simply a matter of implementing a few tools and pulling out some waste (read cost), then most senior leaders would have the appropriate level of commitment. Unfortunately, this is the typical level of understanding and commitment. If the money does not show up immediately, senior leaders quickly lose patience. But lean PD is not as likely to show immediate cost savings. As we have seen the power of lean is in changing the basic structure of the management of people, processes and technology

and thus a shift to a new way. A new way of managing is not finished in two to three years and cannot be delegated to a middle-management level lean department.

Although each of these truths represents a significant challenge, the organizations the authors have worked with have made improvements that strongly suggest these challenges can be confronted and overcome and that results are well worth the time and energy involved. One reason for this is that most of these product development organizations began the journey because they recognized they had to make fundamental improvements to a process that was seriously out of control. A second, and even more important reason, is that once the results start to come in these organizations have begun to recognize that the methods and techniques utilized during *kaikaku* are the beginning of a journey, not an end goal.

Develop an Internal Change Agent

In *Lean Thinking*, Jim Womack advises that the first step on the path to a lean revolution is to find an internal change agent. We agree. It is important that someone in the organization truly own this effort—someone who has the respect of the organization, is unrelenting, and is perhaps even a bit tyrannical, and who will drive the change process through the inevitable difficulties it will encounter. Although this person must understand the development process in your organization, he or she need *not* be an expert in lean product development. Such expertise is helpful, but it is more important for this person to believe passionately in the need for change and be committed to learning. The suitable internal change agent must have appropriate organizational rank, authority, and credibility to make things happen, as well as the unflinching support of the most senior leadership in the organization. Furthermore, this individual must be accountable for the results, with tangible, time-bound deliverables. With these characteristics and an excellent teacher, your internal lean change agent can be developed.

Get the Knowledge You Need

Changing a PD system is a complex undertaking, and your change agent or senior leadership should seek out an experienced and knowledgeable teacher. It is advisable, of course, to get someone with first-hand experience in lean product development and who has experience in changing a product development system. (Someone retired from Toyota is one possibility for this role.) Such individuals are a relatively rare commodity. Indeed, there are many more people with commensurate experience in lean manufacturing, than those possessing the combination of lean product development and lean transformation experience we describe. In fact, it is nearly an empty set because so few PD organizations have transformed to lean. Given the choice between a change agent who knows something about the theories of lean product development and someone with actual experience in a lean PD system, your best choice is the latter. In some cases a skilled lean manufacturing change agent can be helpful but it takes dedication to learning about lean product development, which as we have noted, is different from TPS.

Identify Manageable Work Streams to Understand PD as a Process

As we have said, we advocate that you start your transformation with the PD process and let the work drive your lean PD system. But, you can't work on something you can't see. In most organizations, the PD process is long, complex, and poorly understood. To see the process you must understand the main tasks required to bring a product from concept design into manufacturing, as well as the sequence of those tasks. This is the product-development value stream. The challenge is coming to grips with a process that is typically a long iterative endeavor, involving many diffuse technical disciplines, and thousands of individual steps. But as W. Edwards Deming taught:

> "If you can't describe what you're doing as a process, you don't know what you're doing."

We described PDVSM in Chapter 17 as a tool to help understand your PD process. But where do you start? At a high level you will have very gross activities such as concept design, engineering, and testing. How can we take action on such macro-level activities? To get your mind around product development as a process it can be very useful to move down one level and divide the product development value stream into a number of smaller individual *work streams* that experienced, technically knowledgeable teams can work on.

Work streams are usually major steps within a process, such as prototyping or parts procurement; they can even be the complete development of a specific product subsystem, such as a door-assembly development work stream in the case of automobiles. Often these are boxes in what we call the macro-level value stream—the highest 30,000 foot level. As you understand at this level where the waste is and where you want to go with your future state you can then pick out individual process boxes and develop full-blown value stream maps for these processes. We call these work streams. Each work stream should have its own cross-functional teams, its own more detailed current state map, a detailed future state vision, current state and target metrics, and an action plan. In fact, in many cases, we do not even develop an action plan at the level of the first macro-level value stream map. Rather the work plans are at the work stream level. Of course if the product is simpler and the organization relatively small a single map may be sufficient (as illustrated by the People Flo appendix).

By decomposing the entire PD value stream into these individual work streams, you can make the process much more manageable. Simultaneously, you will be engaging the deep technical knowledge necessary to recognize true opportunities and make decisions about potential process improvement enablers. In this way, technical specialists can focus on the area of the process that they understand best and enlist the support of their home organizations, thus bringing core functional organizations into the effort.

One example of how to organize this effort is to organize your *work stream teams* according to the various functional organizations involved in the PD process. For example, you might have a product planning work stream team, a concept development work stream team, a prototype development work stream team, etc., all organized around a specific product or product family. There are many possible ways to organize your efforts, depending on your product development value stream. The important point is that you need to identify and organize around the critical work streams that make up your PD value stream. In a large organization, you should assign the work-stream team leaders to this effort full time; they form the nucleus of the change team reporting directly to the change agent. You should also maintain strong links between each of the team leaders and their home functional organization, with the full support of that organization's line management.

Integration Mechanisms (*Obeya*/Design Reviews)

We have found that it is best to treat your change effort like a new product program in a lean PD system and your change team as a lean PD project team. In this way, you start to establish norms and practices and lead the effort by example. It is also important to integrate your change team and not leave members to work on their respective work streams in isolation. One of the best ways to do this is to collocate the team in an *obeya*, the same way Toyota collocates its product team leadership, and uses it as a lean product development strategy room. You should assign a section of the *obeya* walls to each of the work stream teams. Here, they can post the latest information and share value stream maps, A3s, pilot and learning trial status, and other relevant initiative metrics across teams. Each of the individual work stream teams can use the room as a meeting location where they can learn what the other teams are working on. This informal cross-team communication should be supplemented and supported by formal meetings at least weekly to give team leaders the opportunity to review each other's value stream maps, A3 progress, and process sheets in a structured disciplined way that enables them to provide valuable input into each other's efforts. Because of the interdependency of most tasks in product development, this "socialization" process among the teams is critical. You must always remember that you are working to create an integrated product-development value stream—not isolated work streams that lead to optimizing local process at the expense of the greater process. Only in this way can you and the individual work stream teams truly understand the challenges of the current state and collaborate on creating the future state.

Enrollment of the Line Organization

Your line organizations must own your conversion to a lean PD system— by this we mean those with operational responsibility for getting out the product designs such as director of engineering and managers of engineering departments. Do not relegate it to the status of a staff initiative. *This requires senior leadership to make career commitments to the effort and back it up by supplying the required resources and selecting the best people to lead it.* Leadership must send a message that they are serious about the transformation process. In addition, functional line leadership must own the development and execution of the strategy. One way to accomplish this is by creating *implementation teams* tasked with the execution of the changes identified by the work stream teams. This must be one of the line leader's primary objectives, and his or her career success should be tied to the success of the effort. Remember that there is nothing more important to any business than product, and the success of your lean product development conversion may determine the very survival of the enterprise.

Start with Your Customer

There is no way around this prerequisite and you have to get it right. Take the time and effort to truly understand the nature of your market, your competitors, and especially what your customer perceives as value. *The spirit of your customer must be pervasive in your organization—a part of every decision you make—at every level.* To deliver on this in a tangible way you must align the entire organization and manifest customer value throughout. One of the best ways to do this is to develop your own version of the CE paper that spells out the value-creating strategy of each product development project and how each person will contribute to that goal. Identify what this product must be and what it will not be and communicate this consistently. Make certain that your process includes a method for alignment (such as *hoshin kanri*) that will allow all participants to create aligned objectives that are understood at all levels. Make the objectives performance based and measure success. Finally, make your customer a part of every discussion and decision in your organization. Always ask, "What is best for our customer?" And then act on the answer.

Beyond being clear on what your customer needs in new *products*, you need to understand what customers define as value to improve the *process* of product development. The basic tenet of Toyota Quality Management is that every function has a customer. Ultimately, this refers to the buyer and user of the product, but there are many intermediate customers along the value-creating chain. Recall, for example, an example mentioned earlier in this book, which described the relationship between Toyota's body engineers and styling engineers. The styling engineers were interim customers whose designs would attract end customers, and the body engineers' initial thrust was to satisfy the needs of these interim customers by staying true to their vision.

One final recommendation is to be certain to make your CE and your project team powerful customer advocates within your organization. They must bring intimate knowledge of customer-defined value to bear on the product under development and have the respect and organizational clout to make things happen.

Grasp the Current State of Your Lean Product Development Process

It is essential to start your lean conversion with a full and brutally honest understanding of your existing product development process. Toyota refers to "grasp the situation" as the first step of any problem solving or improvement process. Only when you truly understand your current process can you create a future state for that process. And only when you fully grasp the future state process can you begin to make good decisions about organizational structure, roles and responsibilities, required skill sets, and the tools and technologies that will best support the execution of a lean PD process. In other words, *let the needs of the process drive other system requirements.* By doing this, you will drive alignment of people, process, and technology. If you begin reorganization without a deep understanding of the work you need to do, your future vision will be uninformed and unrealistic and you will invite unanticipated future resistance that will likely lead to confusion and churn. This process must be focused, deliberate, and systematic. Value stream mapping is a key tool for facilitating this process, and this begins with a deep and thorough assessment of your current processes, which forms the foundation for developing a future state vision.

If you were to visit a physician who prescribed progressive radical surgery before conducting even a routine examination, you would probably run, not walk, out of the practice or hospital where this occurred. Nonetheless, that is exactly what many intelligent managers do within the context of product development system transformation. They subject their organizations to an ever-changing array of "remedies" without fully understanding the ailment. In both cases, it is clearly preferable to use a more scientific approach that includes precise data collection, rigorous data analysis, meaningful diagnosis, based on experienced judgment, thoughtful prescription of a validated remedy, and thorough follow-up.

Industry research suggests that during the 1980s or 1990s, nearly every large company created a concurrent engineering program of one type or another to improve its PD process. Given this, it is likely that your own company participated in this trend. The solution at that time was to create a "phase-gate model," defining clear phases that ended in gates containing specific requirements before a program could pass through to the next phase. Companies developed standard timing at each gate, and the gates were set up to maintain and monitor these PD processes. Invariably, most companies found there was a need for increasing detail to define these "standardized processes." Today, the authors work with many companies in which leadership truly believes that their people are following these phase-gate models, or that any deviation from the model is the cause of all current ailments. Close examination of the current state of product development in these companies shows that there is usually little resemblance between formal process requirements and what is actually occurring. The premise is flawed, and its first fallacy is that an outside, centralized organization can develop a detailed formal process in isolation, teach it, enforce it, and expect people to follow it. There is simply too little interaction (if any) among the people, process, and technology, to support these fantasy models.

Experience and research show that most companies do not truly understand their current state processes. As a result, they often believe that product development in their companies requires far less time and resources than it really does. As an example, one of the authors worked with a company that assumed minor component design could be delivered eight to twelve weeks from receipt of the master product-design information. However, close examination of historical data from past programs (design storage data from corporate servers, purchase orders, etc.) revealed delivery times that were closer to sixteen weeks, with many requiring more than twenty weeks to complete. Without this, data management was just assuming an arbitrary target and lacked a clear plan for achieving it.

Another aspect in the PD process that companies commonly underestimate is the number and effect of engineering changes. Late engineering changes and the resulting expensive rework that they cause are the number one source of waste in every complex PD process, regardless of industry. Companies commonly underestimate these changes by 50 percent or more.

Still another mystery for many product development organizations is how their engineers spend their time. There has been a great deal of discussion about engineers in North American companies spending significantly less time doing "engineering work" than their counterparts at Toyota. Although the research associated with these claims is not conclusive, the authors' experiences with both product development systems suggest that there is some truth to this. The question that emerges from this is: what is it exactly that these engineers spend their time doing?

A number of the companies assume that their engineers spend most of their time in meetings. However, empirical data the authors collected suggest other time wasters. This data shows that although specific activities vary across companies, engineers in non-lean companies spend a great deal of time on the following:

- administrative tasks, such as checking parts lists or chasing down purchase orders.
- creating nonstandardized development and test plans for their parts and developing customer work arounds to adjust ineffective planning systems.
- providing status information to third-party reporting organizations (usually to be provided to senior leadership).
- filling out forms, populating databases, and other tasks associated with demonstrating compliance with requirements for the benefit of auditing organizations such as Quality Assurance who monitor the core engineering team.

Although there was no significant difference in the time engineers spent in meetings, there was a big difference in the cadence and effectiveness (value added) of the meetings attended by engineers in lean and non-lean systems. Consequently, it can be quite useful to reexamine how you manage meetings, who attends (and who does not), what their purpose is, and the cadence in which they occur.

While it is clear that attending meetings where no one makes a decision and no new information is exchanged is non-value-added time, the distinction between value added or non-value-added engineering activities is less clear. The value of specific creative aspects of certain engineering activities, for example, may be difficult to judge. In many companies where such activities are considered "the natural iterative nature of the work," the result is often shoddy or incomplete engineering that requires rework later in the process. This often results in engineering changes that drive significant delays, expense, and frustration to downstream activities. Firms that do not subscribe to this fuzzy interpretation of creative development work and operate within the strictest definitions of rework are the most productive. A previously cited example can be revisited here to illustrate this point: reworking dies by press grinding to produce an acceptable panel. Truly lean companies (like Toyota) see this as waste and an indication that the die engineering process is flawed. Consequently, they rigorously attack the source of the waste and make "die tryout" a target of process improvement efforts. The bottom line is that lean enterprises have a clear grasp of their real world PD value stream and know-how value is created for their customers, and the specific sources of non-value-added time. A big part of this understanding comes from PDVSM, but you must also look at your current state of the entire system of people, processes, and technology.

When you study your PD process, you should choose experienced people from your core product development groups as your task force leaders because they will provide valuable insights into the current state of your product development value stream, particularly within their respective disciplines. Be aware, however, that much of what they will provide will be anecdotal and incomplete, and because they each come from different disciplines, their stories will often contradict each other. This should not come as a surprise, but it should also not be viewed with dismay. Describing a PD system is a bit like the ancient parable about the three blind men describing an elephant for the first time. Each function sees the system from a unique and limited vantage point. The solution: collect real data. PDVSM is one tool for doing this and organizing the data in a sensible manner.

Driving to Real Culture Change

There is a reason that we showcased PDVSM as a tool in Chapter 17. It is the same reason that we recommend value stream mapping as a starting point for transforming manufacturing processes to lean. The reason is that the value stream is the process of delivering value to your customers—it is your reason for existing. Without high value adding processes there is no reason for your organization. Thus we start with a focus on the customer and the processes that adds value to your customer. PDVSM provides a tool for doing this. It also provides a concrete way to get started on the more difficult problem of changing your culture which is what will make lean PD something real and sustainable.

Culture is the shared values and beliefs of an organization. The key word here is "shared" because cultures vary in strength based on how much is shared. In strong cultures, the organizational population has strong values and beliefs in common. In weak cultures, different people think and believe different things and have little in common. For a lean product development organization to be successful, the best culture includes shared values and beliefs about five things: organizational priorities (what really matters), the way work gets done, the way people communicate with each other, the way problems are solved, and the way decisions are made.

In most traditional product development organizations, the culture is quite weak. There are no strong, commonly held beliefs and values. People are often beaten down and have concluded from their experiences that product development is a chaotic, uncontrollable process in which mistakes inevitably slip through and spending time fixing errors is natural and inevitable. They are often cynical of leadership and its ability to manage the place and leery of any "new program" to improve things.

To create a high-performance PD organization with a strong culture, you will need to remember many of the things discussed in this book and act upon them. The customer comes first. Product development is about creating valuable products for customers. People need to work together in teams toward common goals. Leaders must be technically strong and have valuable experience that they can pass on to junior members. The best way to understand a problem is to go and see first hand (*genchi genbutsu*). Deadlines and targets are to be met. It is always possible to achieve a challenging goal if you try hard enough. No detail is insignificant. Standardized processes are necessary for continuous improvement. And so on. The actions taken in implementing lean processes reinforce these beliefs and values every day. Success breeds success.

Moving from a weak culture means moving away from negative work attitudes and behaviors engrained in your people and moving toward a positive, forward-looking, high-performance culture. It will be a challenging journey, one that will occasionally present tempting detours you should avoid. The first thing many companies think of, for example, is instituting a culture change program alongside of process improvements and technical changes. The goal of such programs is to have technical experts value stream map, eliminate waste, and develop tools while "change management" experts change the culture. Unfortunately, *a full frontal assault to change a company's culture never works*. Culture is subtle and does not respond well to change by directive. Telling people what to think, communicating "new" values and beliefs through slogans and emails, and educating them in the new way does little, if anything. In fact, telling people to change their thinking is more likely to reinforce the very thought patterns you are trying to change. People tend to become affronted or defensive and resistant when approached in this way.

A better way to change culture is by changing the way people do their work. In fact, if you do a good job with the value stream-mapping workshop, you are already making progress in changing culture. Consider the following example. In the current situation, there is a standard process for doing engineering work. Management's myth is that people follow the phase-gate process. When management finds things are not going well, it turns to the formal process of rewards and punishments. Tighten up the deadlines, find people to blame, fire someone, and let everyone know that they need to follow the phase-gate process to the letter. If a company is seriously committed to changing culture, however, a value stream-mapping workshop, immediately challenges some of the old cultural norms by:

1. taking an honest look at the current reality.
2. getting a cross-functional team of people together to share their realities from different perspectives.
3. admitting there is something broken in the current system, and it is not the fault of any one person, and we are focusing the team on the process.
4. letting this team help define the future state by enrolling them in the process.
5. empowering the team to follow up with real action.
6. setting up a cross-functional team to experience success.
7. demanding that top management take this seriously and support real change.

One value stream-mapping workshop is not enough to tip the scales on culture change. But it is a start because you have begun the process of getting people engaged in asking the right questions and have empowered them with a tool that can provide meaningful answers. Of course, more important than the workshop are the changes that follow. To assess the scope and breadth of those changes, you will need to ask yourself other questions: Are there serious efforts by the company to implement the changes? Is there immediate follow up with action items? Are you creating new metrics to track progress? Does senior management continue to take an interest and seriously watch the change first hand? Does management allow enough time and create the right environment for people to work on and implement the improvements? Are you working on winning over converts who can become leaders of the new culture? If you can answer these affirmatively, you are on the right track.

A tool like A3 is also an opportunity to start to drive cultural change, but only if the leaders engage seriously in becoming the models of behavior that they are encouraging in others. Leaders must:

- put the customer before their careers in their decision making.
- demonstrate the same level of discipline and rigor that they expect from the rest of the organization.
- create common objectives and reward the right behaviors.
- focus significant energy, time, and resources to the hiring and development of the people who determine the culture.

In this way, day-by-day and step-by-step, leaders can create the path of culture change. They must have patience and understand that culture change is a bit like the giant flywheel described in Jim Collins's path-breaking book *Good to Great* (HarperCollins Publishers, 2001). In the beginning, it takes a great deal of concentrated force just to achieve one turn of the flywheel. The organization struggles just to get everyone moving in the same direction. However, if you stay at it, pushing consistently in the same direction, little by little, momentum grows until finally the wheel is a whirring blur of seemingly endless energy. As Collins observes, it is never clear (nor does it matter) which push made this happen—or what single action created the now powerful culture—it was the sum of all that came before it.

People: The Heart of the Lean Product Development System

A company is made up of people, but as a common aphorism tells us, people are not your greatest asset—the best people are. When applied to product development, this means having people with excellent technical skill sets. Companies that want excellent people can make them or find them. Toyota works at doing both—it invests in a rigorous selection process and then spends years teaching employees the Toyota Way through supervised experiences. For companies that have to start with an existing workforce but have some flexibility to bring in new people through expansion and replacement, the six tips presented below can be useful.

1. *Begin at the beginning with your hiring process.* Be certain you understand the characteristics that define a successful PD engineer, and test for those characteristics through a rigorous assessment process. Review your hiring record. Make sure you are selecting only the very best candidates, no matter how long it takes.
2. *Invest in your people.* It takes time to make a great engineer. You cannot rush the process or expect an engineer educated or trained elsewhere to be great in your company. In fact, a strong mentoring

system, time guidelines, and an insistence on demonstrated competencies are crucial. Set up individual development plans for all your people, based on skill-set achievement tied to career progression. Note that developing towering technical competence takes time. Examine your current progression system to determine if engineers are rotating too quickly and broadly to develop deep expertise.

3. *Develop a technical mentoring system.* Establish a technical apprenticeship. A university does not teach the skills that define a great engineer; these skills are learned on the job, working closely with engineers who are already great. Do not assume it will happen automatically. The process needs structure. Assign mentors, assign the right projects, set objectives by time, assess progress, and select the best. Reward teaching and mentoring as a leadership fundamental. Do not promote anyone who does not have the requisite skills in a basic discipline—build a technical meritocracy.

4. *Understand the skill sets that are required.* The skills necessary to be successful in your new, lean PD process may be different from what you currently have. New technologies (such as simulation or virtual reality) may require skills that do not currently exist in your organization. Anticipate these needs and "future" your skill-set requirements.

5. *Augment basic OJT with classroom training.* This is especially true for specific methodologies or techniques (e.g., new design software) that are specific to your company and that can be codified and taught in a classroom setting.

6. *Set up periodic reviews with functional leadership.* Assess the program. Is it delivering the results you need? Is it still current? Are engineers meeting their targets based on the *hoshin* plan. Assess how well engineers are getting input from functional bosses and program managers. To get the best results, partner with HR and functional leadership.

A Roadmap for Lean Transformation

We are often asked for an implementation road map. What is the step-by-step process to implement lean product development? The answer of course is that there is no one roadmap for all companies. This is not something you can implement like a piece of software. It is an organic, evolving process much as it evolved over decades at Toyota. The closest thing we could develop as a road map is some general directional advice, summarized in Figure 18-1. It is shown as a set of discrete phases, which is already an oversimplification as in reality these phases will overlap. Also, they are shown as a linear sequence and each company will need its own sequence that makes sense given how things evolve in the change process. But the general process is probably valid:

1. *Initial Preparation (2–4 months)*—Some preparatory work is needed to get the key executives and managers on board, get some help, get people generally aware of what this is about, and set up an *obeya* from which to run the change process. Note that in this initial phase we are not expecting that senior management is committed but rather supportive. Senior management cannot be truly committed since they have not experienced real lean PD. Nor do we expect complete training before getting started but rather some general awareness. Most of the important learning will begin by doing in the implementation phases. Thus, this initial preparation is very important but also something you want to do quickly. One can view it as the front-loading of product development where if you do all the preparation just right, the implementation will go smoothly but that would be a mistake. At this point you have not really begun "designing" your lean PD process. You are just setting the stage. This is more like learning to swim than designing a detailed mechanism. At some point (in months, not years) you have to jump in the water and get your feet wet.

2. *Pilot Lean Processes (Minimum 1 year)*—As discussed earlier we recommend starting with action in the process subsystem. This is where you jump in the water. You need to start with the customer, identify your key workstreams, map the current and future states, and then get to work implementing. Metrics should be specific to the projects to measure the cost, quality, and lead time of your PD projects. In this phase, not everybody will be involved but you should focus on pilot projects. The main objective is to get experience and learn the power of lean PD. The teams involved will get an opportunity to experiment with and experience all of the lean PD tools in the process of actual use and implementation. You need to get some traction, learn from the pilots, and start to develop the cultural change by doing, so there is some momentum to go into the more dramatic transformations to your organization. This will take at least one year. When you have successes

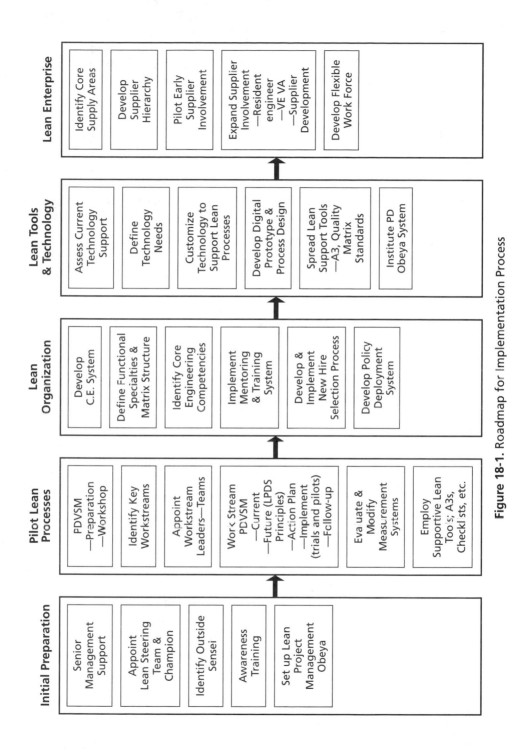

Figure 18-1. Roadmap for Implementation Process

under your belt, some trained internal consultants, and senior management are getting excited about lean as the way forward you are ready for the heavy lifting.

3. *Lean Organization (Years 2–5)*—After getting some experience in implementation and when senior management is more convinced that lean PD is real, applicable, and produces results, you can begin the more arduous task of changing your organization. We should note that a number of the companies with which we have worked never got to this stage. Once they got measurable improvements in the value steam pilots they believed they had mastered lean and doing more value stream mapping and process improvement was all there was to it. They made a serious mistake. They had just gotten started with the easy stuff. Now is the time to start to take seriously the concept of a "chief engineer system." We emphasized that the chief engineer is much more than a traditional project manager or even more than what many companies call "chief engineer." It takes Toyota decades to raise a chief engineer and the chief engineer is part of a broader cultural system that supports this critical role. You do not have decades, but you cannot expect to change business card titles and have a real chief engineer. We suggest starting with a pilot project and picking the person closest to the characteristics we described as the chief engineer at Toyota. Then let that person participate in developing a selection and development process for future chief engineers.

This is also the time to reexamine your organizational structure. You may have a matrix but is it effective? Do you have the right functional departments to give you the core engineering competencies you need? Are the functional departments building strong technical expertise? And do you have a process like policy deployment that aligns engineers toward a common focus on meeting congruent targets?

4. *Lean Tools and Technology (Years 2–Forever)*—You already have made a good start on this in the first year, but in year one you are focused on pilots and you only focused on tools and technology that required little or no capital investment. In the pilots you will discover limitations of your current technology. This provides the starting point of looking at what types of technology will be needed to support true lean PD. Digital prototyping and an effective review process may be something you already have or do not have. It is

essential for any complex product in this day and age. Even if you have it, review it and we guarantee there is great opportunity for improvement—not necessarily in the technology but in the way it is used, how you organize to exploit the information from a strong prototype review, and using the results as part of future checklists. You can now also work on spreading the tools that have worked for you across the organization like A3, quality matrices, product and process standards, and *obeya*.

5. *Lean Enterprise (Years 3–Forever)*—Once you have some internal mastery and some stability to your own development process you can begin to bring in suppliers and even your customer to create a true lean enterprise. Again we recommend piloting with a few key suppliers before spreading broad policies across the board. There is a lot of learning needed to do this right. You need to know enough to be a teacher before you can start to spread these practices to suppliers for example. You need to have stable processes to integrate your customers and suppliers into before they can participate in the enterprise. Another major task at this stage is to develop a flexible pool of resources. Again, until you have your processes standardized and documented it will be difficult to train outside technical resources on the standardized approach.

In summary, this is a process of learning by doing, reflecting, and improving. It is PDCA at all levels. In terms of the LPDS model, we are suggesting that you do some preliminary work to get key leaders and resources on board, dive into the process box using value stream mapping as a guiding tool, and through initial pilot projects, begin to develop your people and tools and technology. Once you have a foothold of some experience and knowledge you can begin the process of planning broader organizational and technological changes. If you continue this process step by step, continually reflecting, understanding the true current reality, and developing a bold but realistic vision for the next step, then you cannot fail.

Leadership and Building in Learning and Continuous Improvement

As the reader has no doubt discovered by now, there is a big difference between management and leadership. Managers plan, organize, control, and generally work through the formal system. Leaders capture the hearts and souls of the people—they make things happen and people want to follow them. For this reason alone, "change management" is little more than an industry myth. You do not manage cultural change; you lead it. Two questions that immediately emerge from this assertion underscore the challenge associated with leading a culture change: If the existing bureaucratic organization is the way it is because there are many managers and few leaders, how can managers suddenly learn to lead? And how can they learn to lead a change to the new lean culture if they have never experienced it and do not understand it in their gut?

The way of the lean leader, especially for a leader in a PD organization, lies in teaching, and one cannot teach what one does not understand. To transform the organization, the leader must change first and do so in ways that are visible. By living the change, showing the way, implementing new work processes, leading cross-functional teams, and using the lean PD tools, people create the opportunity to develop into real transformational leaders. Those who learn to teach in this manner open the doors to building in learning and continuous improvement.

The current business environment is hypercompetitive, and it is not likely that things will calm down any time soon; you cannot afford to think in terms of completing this effort. You must build learning mechanisms into your process, making them part of how you develop products. Furthermore, you will need to establish an oversight group, institutionalize your gains, standardize across PD projects, manage shared resources, and maintain process discipline. Above all, you must use all of these to drive continuous improvement.

The concept of finishing your transformation to lean PD contradicts lean philosophy. Lean is about continuous improvement in adding value to customers. It is a journey of eliminating waste through problem solving and continuous improvement. If the journey stops, continuous improvement stops. When that occurs, an organization can no longer call itself lean. One thing that distinguishes Toyota from other lean enterprises is that the Toyota organization is not only a great teacher but also a great student. Offer an observation or suggestion and it is likely to be seriously considered (or studied) for its potential. On the same note, Toyota has always engaged in the study of feedback, whether from its end customers or the people on its shopfloor. Feedback is a gift to a learning organization and a threat to a rigid bureaucracy. In a lean product development system, it is always accorded attention.

Lean PD is about jumping off into the improvement abyss. It is risky. It is exhilarating. There is no return. We hope our work has inspired you to take the leap and begin your long PD transformation journey.

Applying Value Stream Mapping to a Product Development Process: The PeopleFlo Manufacturing Inc. Case

by Dr. John Drogosz

PEOPLEFLO IS A SMALL START-UP COMPANY that designs and manufactures pumps for chemical, petrochemical, and food processing applications. The company was founded by a team of people committed to applying lean manufacturing philosophy and tools in a unique and comprehensive way. Starting with a "clean slate," PeopleFlo's management team designed a new product line and built every core business process into one interdependent lean enterprise. They developed patented technologies, simplified the product so that it could be flexibly built in lean cells, and met every major timing milestone.

An essential part of the company's lean enterprise vision was to rethink how best to design new products. This meant slashing the typical product development cycle from three to four years in the pump industry down to less than one year. In addition, it meant achieving breakthroughs in design for setup reduction, design for cycle-time reduction, and design for inventory reduction. To achieve these goals, the company started with a clean sheet of paper and used the value stream mapping tool not only to detail the best way to design, validate, and launch production of a new pump product but also to manage the product development process. The first step was to start with a high-level map that showed high-level activities, key decisions, and integration points that were necessary to design a new pump. Figure A-1 shows the major phases of the product development process and its milestones.

Once the high-level process map was defined, each process was mapped in more detail. The tasks and key information flows were delineated and timing was set for each task. Figure A-2 shows the process architecture for the preliminary design phase.

As part of the mapping activity, PeopleFlo incorporated several lean product development concepts into its development process. For example, activities were included that allowed design time to examine multiple

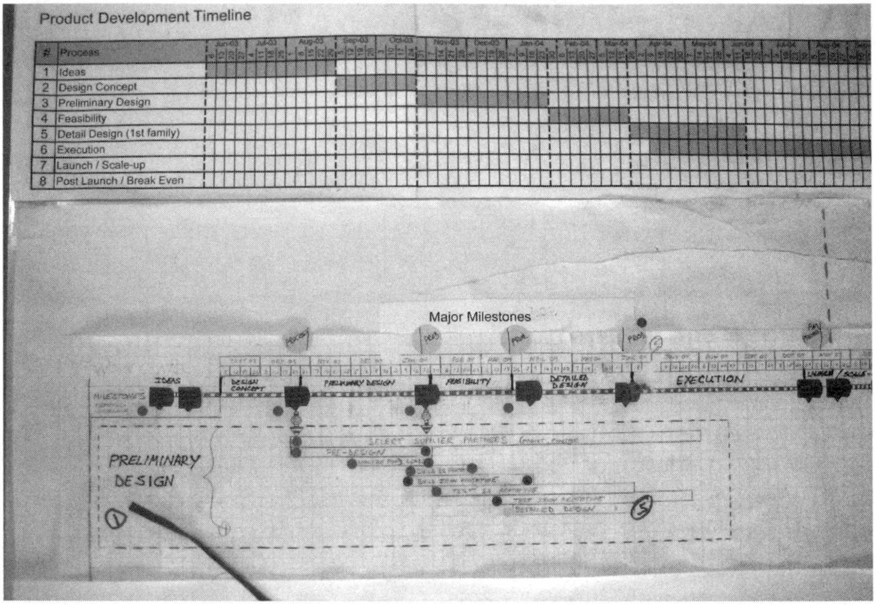

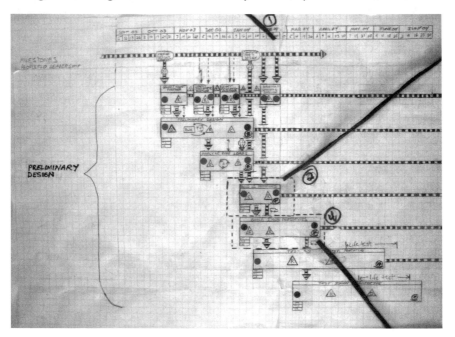

Figure A-1. High Level Product Development Map & Master Time Line

Figure A-2. VSM of Preliminary Design Phase

alternatives (set-based concurrent engineering). In addition, reflection events were hard-coded into the plan to ensure that learning was captured all the way through the development process.

In addition to using the value stream map to define a new PeopleFlo product development process, company leaders used it as a visual program management tool. Figure A-3 illustrates how they used the visual controls to highlight problems/opportunities and indicate program status.

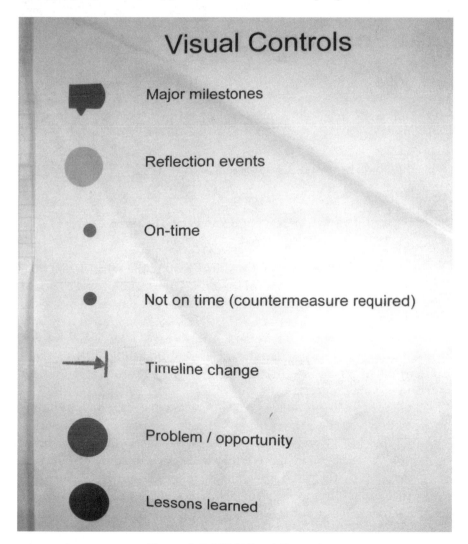

Figure A-3. VSM Visual Controls

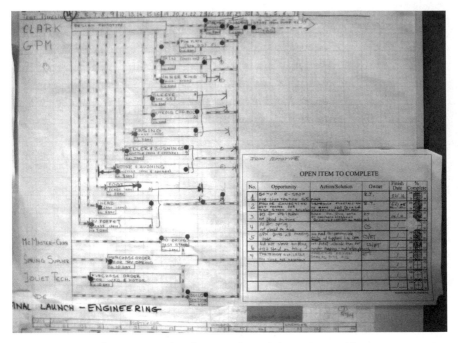

Figure A-4. Visual Controls and Open Issues List

Teams held "morning market" meetings around the value stream map boards to report status. As Figure A-4 illustrates, the program board showed current status regarding timing; red-dot visual indicators were placed directly on the map to indicate where issues were discovered. Any red dots required an immediate countermeasure. An open issues list was also posted at the VSM program board. In addition to acting as a rapid problem-solving tool, the list was also used to capture learning for improving future programs.

As the product development cycle progressed, the team members continued to add new lower-level maps and include tasks and information flows that they had missed in creating the early-stage VSM. By the end of the first product development cycle, they had clearly documented the best way to create their first product family of pumps. Figure A-5 shows the overall process architecture for PeopleFlo's product development process.

The development team is now using the value stream maps from the first family of pumps as the framework for the next family of pumps they are currently designing. The plan is to continue to apply the VSM

Figure A-5. VSM Process Architecture for Pump Family 1

approach to all future product development programs and customize this approach for each product family's specific requirements.

The result has been the development of a complete line of pumps that are in production in less than two years. In that time, PeopleFlo was able to design an innovative new generation product, with 50 percent fewer parts than comparable pumps and develop a unique fixturing system that eliminates changeover time on machining centers. A true one-piece flow cell without any changeover time allows flexible production. One cell has the ability to machine ten different pump parts, then paint, assemble, and ship 80 percent of parts sold, within 24 hours of receiving the order—unheard of within this industry. The cost for a comparable pump made by a conventional competitor was slashed by 50 percent.

The key to the success of this start-up business was a combination of people, process, and technology. Success factors included starting with a clean slate, assembling the right team of employees and business partners, and building one interdependent design-manufacturing system, with one set of goals, based on lean philosophy and tools.

Chapter 1

Womack, James P., and Jones, Daniel T. (1991), *The Machine That Changed the World: The Story of Lean Production*. New York: Harper Perennial.

Morgan, James M. (2002), *High Performance Product Development; A Systems Approach to a Lean Product Development Process*, The University of Michigan, Ann Arbor MI.

Liker, Jeffrey K. (2004), *The Toyota Way: 14 Management Principles from the World's Greatest Manufacturer*. New York, McGraw-Hill.

Taylor, James C. and Felten, David F. (1993), *Performance By Design*, Prentice-Hall, Englewood Cliffs, NJ.

Nadler, David and Tushman, Michael L. (1997), *Competing By Design*, Oxford University Press, New York, NY.

Chapter 2

Morgan, James M. (2002), *High Performance Product Development; A Systems Approach to a Lean Product Development Process*, The University of Michigan, Ann Arbor MI.

Liker, Jeffrey K. (2004), T*he Toyota Way: 14 Management Principles from the World's Greatest Manufacturer*. New York, McGraw-Hill.

Chapter 4

Cusumano, Michael A., and Nobeoka, Kentaro (1998), *Thinking Beyond Lean*, The Free Press, New York, NY.

Ward, Allen C., Sobek, Durward K., II, Cristiano, John J., and Liker, Jeffrey K. (1995), "Toyota, Concurrent Engineering, and Set-Based Design," in Liker, et al., eds. *Engineered in Japan*, Oxford Press, New York; pp. 192–216.

Chapter 5

Liker, Jeffrey K. (2004), *The Toyota Way:14 Management Principles from the World's Greatest Manufacturer*. New York, McGraw-Hill.

Rother, Mike, and Shook, John. (1998), *Learning to See*, The Lean Enterprise Institute, Brookline, MA.

Adler, Paul S., Mandelbaum, Avi, Nguyen, Vien, and Schwerer, Elizabeth (1996), "Getting the Most out of Your Product Development Process," *Harvard Business Review*, Mar.–Apr. 1996, vol. 74, no. 2; pp. 134–151.

Reinertsen, Donald G. (1997), *Managing the Design Factory*, The Free Press, New York, NY.

Hopp, Wallace J., Spearman, Mark L. (1996), *Factory Physics*, Irwin, Chicago, IL.

Morgan, James M. (2002), *High Performance Product Development; A Systems Approach to a Lean Product Development Process*, The University of Michigan, Ann Arbor MI.

Loch, Christoph H. and Terwiesch, Christian (1999), "Accelerating the Process of Engineering Change Orders: Capacity and Congestion Effects," *Journal of Product Innovation Management*, Apr. 1999, vol. 16, no. 2.

Cusumano, Michael A., and Nobeoka, Kentaro (1998), *Thinking Beyond Lean*, The Free Press, New York, NY.

Chapter 6

Kramp, Eric E. (2001), *How soft issues influence hard work, loyalty and a sense of pride to build superior products at Toyota*, Ford Motor Company Internal Presentation, Dearborn, MI, 6 Dec.

Chapter 7

Liker, Jeffrey K. (2004), T*he Toyota Way: 14 Management Principles from the World's Greatest Manufacturer.* New York, McGraw-Hill.

Itazaki, Hideshi (1999), "The *Prius* that Shook the World: How Toyota Developed the World's First Mass-Production Hybrid Vehicle," Tokyo, Japan—The Kikkan Kogyo Shimbun, Ltd. (translated by A. Yamada and M. Ishidawa).

Sobek, Durward K., II (1997), *Principles that Shape Product Development Systems: A Toyota-Chrysler Comparison*, UMI Dissertation Services, Ann Arbor, MI.

Chapter 8

Sobek, Durward K., II (1997), *Principles that Shape Product Development Systems: A Toyota-Chrysler Comparison*, UMI Dissertation Services, Ann Arbor, MI.

Cusumano, Michael A., and Nobeoka, Kentaro (1998), *Thinking Beyond Lean*, The Free Press, New York, NY.

Chapter 9

Rich, Ben R. and Janos, Leo (1994), *Skunk Works*, Little, Brown and Company, New York, NY.

Hammett, Patrick C., Wahl, Shannon M., and Baron, Jay S. (1999), "Using Flexible Criteria to Improve Manufacturing Validation During Product Development," *Concurrent Engineering: Research and Applications*, Dec. 1999, vol. 7, no. 4; pp. 309–318.

Liker, Jeffrey K. (2004), *The Toyota Way: 14 Management Principles from the World's Greatest Manufacturer*. New York, McGraw-Hill.

Chapter 10

Kamath, Rajan R. and Liker, Jeffrey K. (1994), "A Second Look at Japanese Product Development," *Harvard Business Review*, Nov.–Dec., vol. 72. no. 6; pp. 154–170.

Chapter 11

Ward, Allen C., Liker, Jeffrey K., Cristiano, John J., and Sobek, II, Durward K. (1995), "The Second Toyota Paradox: How Delaying Decisions can make Better Cars Faster," *Sloan Management Review*, vol. 36, no. 3; pp. 43–61.

Senge, Peter M. (1990), *The Fifth Discipline*, Doubleday/Currency, New York, NY.

Drucker, Peter F. (1998), "The Coming of the New Organization," in *Harvard Business Review on Knowledge Management*, Harvard Business School Press, Boston, MA; pp. 1–20.

Nonaka, Ikujiro and Takeuchi, Horotaka (1995), The Knowledge-Creating Company: How Japanese Companies Create the Dynamics of Innovation, Oxford University Press, New York, NY.

Kogut, Bruce, and Zander, Udu (1992), "Knowledge of the Firm, Combinative Capabilities, and the Replication of Technology," *Organization Science*, Aug. 1992, vol. 3, no. 2; pp. 383–397.

Conner, Kathleen R., and Prahalad, C. K. (1996), "A resource-based theory of the firm: Knowledge versus opportunism," *Organization Science*, vol. 7, no. 5; pp. 477–501.

Argyris, Chris (1998), "Teaching Smart People How to Learn," in *Harvard Business Review on Knowledge Management*, Harvard Business School Press, Boston, MA; pp. 81–108.

Garvin, David A. (2000), *Learning in Action*, Harvard Business School Press, Boston, MA.

Nelson, Richard R., and Winter, Sidney G. (1982), *An Evolutionary Theory of Economic Change*, Belknap Press, Cambridge, MA.

Pfeffer, Jeffrey and Sutton, Robert I. (2000), *The Knowing-Doing Gap*, Harvard Business School Press, Boston, MA.

Dyer, Jeffrey H., Nobeoka, Kentaro (1998), "Creating and Managing a High Performance Knowledge-Sharing Network: The Toyota Case," *Strategic Management Journal*, vol. 21, no. 3; pp. 345–367.

Morgan, James M. (2002), *High Performance Product Development; A Systems Approach to a Lean Product Development Process*, The University of Michigan, Ann Arbor MI.

Hann, D. (1999), "*Organizational Forgetting*," unpublished study, Harvard Business School, Boston, MA.

Chapter 12

Liker, Jeffrey K. (2004), *The Toyota Way: 14 Management Principles from the World's Greatest Manufacturer*. New York, McGraw-Hill.

Chapter 13

Clark, Kim B., and Fujimoto, Takahiro (1991), *Product Development Performance: Strategy, Organization, and Management in the World Auto Industry*, Harvard Business School Press, Boston, MA.

Chapter 14

Sobek, Durward K., II (1997), *Principles that Shape Product Development Systems: A Toyota-Chrysler Comparison*, UMI Dissertation Services, Ann Arbor, MI.

Morgan, James M. (2002), *High Performance Product Development; A Systems Approach to a Lean Product Development Process*, The University of Michigan, Ann Arbor MI.

Chapter 16

Wheelwright, Steven C. and Clark, Kim B. (1992), *Revolutionizing Product Development*, The Free Press, NY.

Clark, Kim B., and Fujimoto, Takahiro (1991), *Product Development Performance: Strategy, Organization, and Management in the World Auto Industry*, Harvard Business School Press, Boston, MA.

Womack, James P. and Jones, Daniel T. (1996), *Lean Thinking*, Simon and Schuster, New York, NY.

Hopp, Wallace J., Spearman, Mark L. (1996), *Factory Physics*, Irwin, Chicago, IL.

Rother, Mike, and Shook, John (1998), *Learning to See*, The Lean Enterprise Institute, Brookline, MA.

Adler, Paul S., Mandelbaum, Avi, Nguyen, Vien, and Schwerer, Elizabeth (1996), "Getting the Most out of Your Product Development Process," *Harvard Business Review*, Mar.–Apr., vol. 74, no. 2; pp. 134–151.

Loch, Christoph H. and Terwiesch, Christian (1999), "Accelerating the Process of Engineering Change Orders: Capacity and Congestion Effects," *Journal of Product Innovation Management*, Apr., vol. 16, no. 2.

Chapter 17

Rother, Mike, and Shook, John (1998), *Learning to See*, The Lean Enterprise Institute, Brookline, MA.

Morgan, James M. (2002), *High Performance Product Development; A Systems Approach to a Lean Product Development Process*, The University of Michigan, Ann Arbor MI.

Morgan, James M., *Learning to See Product Development*, The Lean Enterprise Institute, Brookline, MA.

Collins, Jim (2001), *Good to Great*, HarperCollins Publisher, New York, NY.

Index

Index

Index

Index

Index

Dr. James M. Morgan has more than 24 years experience in automotive product development and operations management, including almost 20 years at TDM; a tier one automotive supplier of engineering services, tools, dies and vehicle subsystems where he was vice president. He holds MS and Ph.D. degrees in Engineering from the University of Michigan where he completed a three-year, Shingo Award-winning comparative study of Toyota and a North American competitor's product development systems.

Dr. Morgan's research has lead to the creation of a coherent systems model of lean product development which he has utilized in analyzing and improving the development of new products in several fortune fifty companies in both the United States and Europe. Dr. Morgan has published a number of articles and developed and taught classes and seminars at the University of Michigan, M.I.T., the Lean Enterprise Institute, the Lean Enterprise Academy, and the Society of Automotive Engineers.

Dr. Morgan is currently an engineering director at Ford Motor Company.

Dr. Jeffrey K. Liker is Professor of Industrial and Operations Engineering at the University of Michigan. Dr. Liker has authored or co-authored over 70 published articles and book chapters and seven books. He is author of the international best-seller, *The Toyota Way: 14 Management Principles from the World's Greatest Manufacturer* (McGraw Hill, 2004), which speaks to the underlying philosophy and principles that drive Toyota's quality and efficiency-obsessed culture. The companion (with David Meier) *Toyota Way Fieldbook* (McGraw Hill, 2005), details how companies can learn from the Toyota Way principles. He is also the Editor of *Becoming Lean: Inside Stories of U.S. Manufacturers* (Productivity Press, 1997), and winner of the 1998 Shingo prize (for excellence in manufacturing research). He has also won Shingo prizes for his research in 1995, 1996, and 1997. Other books by Dr. Liker include *Engineered in Japan* (Oxford University Press, 1995); *Concurrent Engineering Effectiveness: Integrating Product Development Across Organizations* (Hanser-Gardner, 1997), and *Remade in America: Transplanting and Transforming Japanese Manufacturing Methods* (Oxford University Press, 1999). He is active as a keynote speaker, speaker for executive retreats, and lean consultant, independently and through a company he cofounded—Optiprise, Inc.